SUSTA

Chris Goodall

ALL THAT MATTERS

ALL THAT MATTERS

Also available in ebook

Contents

1

Depleting
natural capital

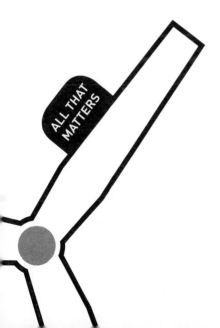

▶ Aquifers

The vast Ogallala underground aquifer stretches across most of the US High Plains, providing water for crops and people from Texas to as far north as South Dakota (Figure 1.1). The irrigated corn, soy and cotton grown across this region provide almost a quarter of America's harvest. This may not continue: in many places across the region water is being extracted from the aquifer at more than 100 times the rate of natural replenishment. As a result, the water table under the ground is dropping sharply and in some places has completely disappeared. Most of the High Plains area is naturally dry, getting far less rainfall than necessary to sustain crops such as corn when grown using conventional agricultural techniques. Future climate change is likely to reduce the amount of rain, exacerbating the shortage of water.

The over-extraction of water for irrigation is threatening the US's future ability to produce food from its breadbasket. The problem is getting worse as expanding cities in the southern US increase their use of the Ogallala aquifer, further threatening the ability of farmers to irrigate their crops, and helping to drive up the price of food. The aquifer is part of Earth's 'natural capital', an endowment of a valuable resource that is being rapidly depleted by the current generation to the probable detriment of the people of the future.

Overuse of shared water resources occurs across the world. In China, the rapidly growing demands of Beijing and other mega-cities are rapidly running down the water stored in the aquifers of surrounding regions.

▼ Figure 1.1 Ogallala aquifer

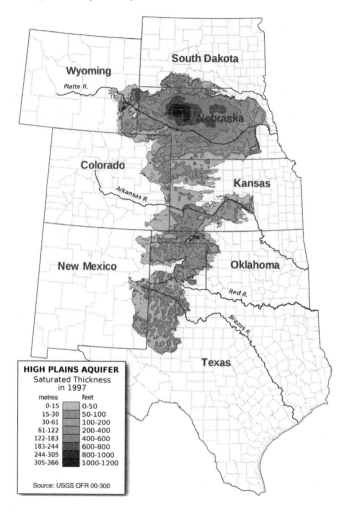

HIGH PLAINS AQUIFER
Saturated Thickness
in 1997

metres	feet
0-15	0-50
15-30	50-100
30-61	100-200
61-122	200-400
122-183	400-600
183-244	600-800
244-305	800-1000
305-366	1000-1200

Source: USGS OFR 00-300

ALL THAT MATTERS: SUSTAINABILITY

The Hebei Plain around Beijing can annually provide about 7 billion cubic metres of fresh water to urban areas without depletion of the aquifer, but is currently producing almost twice this figure, with further increases year-on-year. Groundwater is being depleted rapidly, with scientists forecasting that the water table will drop 40 metres by 2030. The water that is pumped from underground is increasingly polluted. Some parts of the aquifer have dried up, causing land subsidence, falling agricultural production and local water shortages.

Many countries in the Middle East are also rapidly running down their reserves. Rulers have traditionally bought social peace by providing abundant fresh supplies but the consequences are becoming more obvious every year. By 2040 Abu Dhabi will have completely used up all its groundwater supplies – which took many millennia to accumulate – if it does not moderate its wasteful use of water. Desalination is at best a partial solution in the Middle East; the production of fresh water from the sea is highly energy intensive and produces salts that cannot be easily stored. Throwing the waste back in the sea increases local salinity, affecting fishing stocks and changing the coastline ecology.

Although we all find it difficult to define the word 'sustainability', we can immediately see when today's exploitation of our natural world is threatening the future, or causing clashes between the lucky and the disadvantaged. The determined and deliberate overuse of 'fossil' water, which in many places was laid down in natural aquifers at the end of the last Ice Age, is perhaps the most obvious example everywhere across the globe.

In the High Plains of the US, the hinterlands of Chinese cities, the rich and poor countries of the Middle East, and in numerous other regions, today's users are running down underground reserves and in so doing are affecting the life chances of future generations. Disputes between countries and between urban dwellers and farmers in Texas or central China are exacerbated by the abiding sense that one group is profiting by exploiting limited resources not available to others.

▶ Reversing an aquifer's decline

Humankind recognizes the foolishness of the destruction of vital aquifers but action to protect underground water supplies is rare. To be successful, the restoration needs a strong authority, such as a government, able to ban the drilling of wells or put such a high price on water that farmers and householders carefully ration their use. One successful example is the improvement of the Sparta aquifer across the boundary of the US states of Arkansas and Louisiana. In the mid-1990s water levels were dropping, threatening supplies to homes, farms and businesses. A greater threat was the destruction of the geology of the aquifer itself: as water was pumped out, the weight of surrounding rock threatened to crush the porous structure of the layer containing the water. The loss of the aquifer would have been irreversible. Additionally, salt and other minerals might have polluted surrounding water supplies.

One calculation suggested that extraction needed to fall by almost three-quarters and action was taken at the end of the 1990s. Industries needing large volumes began to use groundwater in rivers rather than from the underground source. Other users began to be charged for their water and successful efforts were made to reduce water consumption. Water levels in many of the wells that tap the aquifer have now risen. The aquifer is still well below the level of the early twentieth century but the immediate threat to fresh-water supply has lifted.

▶ Overfishing

In the case of aquifers and other natural assets that can be freely exploited by many, no individual has a substantial incentive to maintain the long-run viability of the shared resource. You might try to restrict your use in order to conserve a valuable resource but others will still take what they can. This phenomenon is often called 'the tragedy of the commons' and the most obvious example is the overfishing of the oceans. Since no one owns or controls most of the open sea, ruthless pillaging has destroyed some of the world's most productive fisheries. Fewer and fewer fish survive to reproduce, cutting supplies now and in the future.

Despite advances in the technology for finding and netting fish, the total weight taken from the world's oceans is now less than it was in 1990. Current fishing patterns – using larger ships, bigger nets and focusing on areas not yet fished – suggest that the slow worldwide decline in ocean catches will continue and probably accelerate.

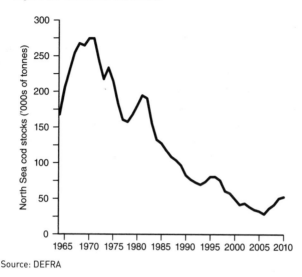

▼ Figure 1.2 North Sea cod stocks

Source: DEFRA

The North Atlantic fisheries have been particularly badly affected, with most species suffering severe depletion. Estimates of the total weight of cod in the North Sea shows an all-too-typical pattern over the last 50 years (Figure 1.2).

Across the Atlantic, even worse problems hit cod stocks off Canada (Figure 1.3). Huge factory trawlers overfished the northern cod and almost completely cleared the adult species from the sea in the 1990s. The smaller fish on which the cod had fed increased rapidly in number and size. One study showed a nine times expansion in the total mass of the predator-free fish. These newly numerous fish ate the remaining baby cod as well as unsustainably large volumes of the

ALL THAT MATTERS: SUSTAINABILITY

▼ Figure 1.3 Decline of Atlantic cod stocks

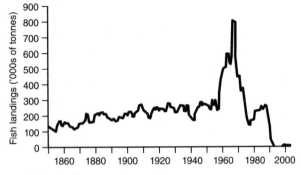

Source: Millennium Ecosystem Assessment

zooplankton on which they normally feed. The effects cascaded further as phytoplankton, the food source for zooplankton, became far more prevalent as their predator disappeared. The phytoplankton consumed more oceanic nitrogen, producing blooms on the sea surface that reduced the availability of oxygen to other marine creatures. It was not just the loss of the cod that mattered; it was the effect across the whole of the marine ecosystem. The disappearance of the predator triggered large, and wholly unpredicted, changes in biodiversity.

Twenty years later these ecological effects are very gradually unwinding. Cod can now be found off Newfoundland again but the total number of these fish is probably less than 10 per cent of the level of the 1980s. Even at this level, local fishermen continue to put pressure on the Canadian government to allow the rebirth of the cod fishing industry. Understandably, long-term sustainability often comes a remote second to the need to earn money this month.

Allowing fish stocks to rebuild in order to increase the weight that can sustainably be harvested in the future is proving a worryingly difficult problem all around the world. It seems simple enough – all we need to do is work out the maximum weight of a particular fish that an area of the ocean can support and then calculate the amount that can be harvested each year without running down the stock. For any individual fish, scientists can provide estimates of these values with a few months' work. Nevertheless, severe overexploitation continues, reducing future yields of fish to well below these optimal levels. As one scientist once commented, 'If we can't solve the overfishing problem, what hope do we have of ever getting to grips with really difficult issues such as climate change?'

Some countries have shown that fish stocks can be carefully rebuilt. The summer flounder or fluke is an important species for both commercial and recreational fishing off the Atlantic states of the US. Harvests fell catastrophically in the late 1980s as a result of previous over-extraction. After several failed attempts to reduce the catch, fishing regulators eventually succeeded in enforcing strict limits on the weight of fish landed. By the summer of 2011 the number of summer flounder had rebuilt to a level that scientists said was adequate and the limit on catches was relaxed. Carefully enforced laws, rigorous science and a willingness to stand up to those who want to harvest too much today can produce sustainability.

▶ Transferring the costs of our consumption

Fresh water and fish are two examples of one generation's willingness to run down the world's natural capital at a severe cost to the people of the future. Examples of this behaviour abound, and not just in our time. As Jared Diamond points out in his highly influential book, *Collapse,* civilizations as diverse as the Anasazi of New Mexico, Easter Island and the Central American Mayan culture seriously abused the ecological systems on which they relied, and suffered – sometimes to the point of catastrophe – as a result.

Many separate civilizations, each covering a tiny fraction of Earth's surface, have destroyed themselves by their unsustainable behaviour and abuse of the local environment. Life elsewhere has continued normally. Humankind's technical progress has now created a more general threat, or rather series of threats. The resources of oil, gas and coal in the Earth's crust, used in ever-increasing volumes around the world, have given us domination of the entire planet and a remarkable increase in material prosperity. Unsustainable behaviours arising from the running down of natural capital can now take place on a planetary scale.

As a way of illustrating our implicit decision to take easy prosperity today and to ignore the future, George Monbiot's book *Heat* uses the powerful analogy of the legend of Faust, a person who traded a period of

24 years of worldly satisfactions for eternal damnation. Humankind, Monbiot says, appears to have entered into a similar deal with the planet that it inhabits. And, like Faust, some of the effects may be irreversible. Depleted aquifers may cease to function, fish stocks not return and greenhouse gas increases change our climate for millennia to come, even if we stop emitting CO_2 today. On current trends, we face at least a 3 to 4-degree increase in temperature by the end of the century, increasingly dangerous floods and droughts, and a potentially disastrous rise in sea levels.

▶ Sustainability

Now is a good moment to say what this book is about. Two hundred years ago most of humankind lived a lifestyle that was sustainable. We used the biological production of the land in the form of food from animals and from crops, wood for fuel and for building, fibres such as linen and wool for clothing. Humankind quarried small amounts of stone for building and mined limited amounts of iron ore for the metal items most useful for everyday life. Lifestyle then may have been sustainable, in the sense of being able to continue for ever, but it certainly wasn't comfortable for any but the very rich. Food supply was unreliable, lives were short and disease an ever-present threat. Backbreaking work was the daily norm.

The exploitation of fossil fuels, starting in the UK and spreading around the world, has given us the ability to provide food for billions of people, machines that do all our heavy work, and vital conveniences such as electricity for lighting and natural gas for heating. It has allowed us to devote resources to improving lives through the application of science. Our buildings use steel that has been smelted using coal. Plastics made from oil give us indestructible packaging to protect our food as it is shipped hundreds of miles in diesel-fuelled trucks.

Virtually nobody wants to return to the days before we had almost limitless supplies of cheap energy. Whatever you may think of today's oil prices, they are still a tiny fraction of how much it would cost to hire human beings to do the same amount of work. A manual labourer working for £10 an hour, and without help from fossil energy, might provide 5 kilowatt hours of useful energy in a working day. This amount of energy is contained in half a litre of petrol, costing less than a pound in Britain or a dollar in the US.

The sustainability challenge can be simply expressed. If current forecasts are accurate, world population will peak at about 10 billion in 2050 before declining, probably sharply. (After this point sustainability will become less of a challenge because resource needs will have started to fall.) Can we offer all the 10 billion the prospect of secure food and water supply, abundant supplies of energy, reasonable clothing and shelter, and access to telecommunications and computing? Are we able to do this without dangerous loss of biodiversity or catastrophic pollution of land, sea and atmosphere? In other words, are we able to create a world where the conditions in which the most prosperous 1 or 2 billion live today have been extended to everybody in a much larger global population at mid-century?

This is the most difficult challenge the world has ever faced. There is no doubt it is technically feasible – the progress of science means we are getting better and better at creating sustainable solutions – but I sometimes doubt whether we have the far-sightedness to wrench the world away from the exploitative habits of the past century or more. This book is about where sustainability is easy to achieve, where it is difficult but feasible, and, most important, where it is impossible without painful changes in lifestyle.

It is also a plea that the world accepts that new technology has a vital role in giving us the advances that make long-term prosperity possible. The argument for sustainability is simply an ethical one – it is wrong to inflict the severe problems caused by us onto future generations – and solutions are to be found in science and in engineering. Sustainability is about calculating the limits humankind has to live within, and then using our scientific genius to give us all a good life within those boundaries. To paraphrase David MacKay, author of a highly influential book on sustainable energy, we need numbers, not adjectives, when assessing the severity of the problem and its possible solutions. I hope you will find some of the crucial numbers in this book.

2

Are we going to run out of anything?

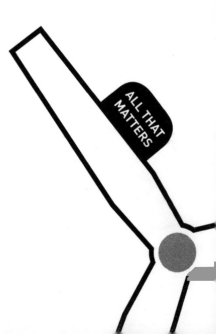

ALL THAT MATTERS

The achievement of sustainability requires us to meet two deceptively simple conditions:

1 The world's current use of minerals, fuels and the products of the soil must not reduce the resources available to future generations if this constrains their standard of living to below ours.

2 Our management of the planet must avoid pollution, disruption or degradation that makes it substantially more difficult for future generations to match our material prosperity, health or pleasure in our physical surroundings.

This chapter is largely about the first of these two conditions: Are we likely to run out of raw materials?

The Earth's crust provides humankind with just three broad categories of supplies:

1 **Fossil fuels.** These provide the energy to give us protection from heat and cold, meet our transport needs and allow us to escape physical drudgery by using machines. Much smaller quantities are used to make plastic and fertilizers. In the developed world fossil fuels provide us each with roughly 120 kilowatt hours a day of useful energy in the form of electricity, natural gas and refined oil products such as petrol. (To help put this in context, the typical European house uses about 10 kilowatt hours of electricity a day.) We use about three or four times as much energy as before the Industrial Revolution.

2 **Minerals.** The most important minerals are the ores from which we make metals. Sand, gravel and limestone, the most important ingredient of cement, give us the materials to make buildings.

3 **Biomass.** This is the word for everything created, directly or indirectly, by the process of photosynthesis. This covers all our sources of food, wood for fuel, paper and construction, and fibres such as cotton and wool. Biomass is also increasingly used to make liquid fuels for transport. Carbohydrates such as corn are used to make a petrol substitute while food oils, made from such things as palm nuts, are used to replace diesel. To give us a sense of scale of our use of biomass, a human being needs about 2 kilowatt hours of food energy each day, less than 2 per cent of the energy we get from fossil fuels. Of course, we also need fresh water, for agriculture and for our own needs.

Perhaps, like me, you find it difficult to believe that all our needs can be reduced to these three classes of materials. If so, just think through the things you have used in the last day. Food comes from the sun, either directly in the form of plants or indirectly from animals or fish that have eaten plants. Plastics are made from oil, our housing is largely made from cement, steel and wood, and our clothes either from fossil fuels or from cotton from the fields.

▶ Fossil fuels

We hear lots of talk about 'peak oil', suggesting we are rapidly running out of energy. It is certainly proving difficult to find new large oil fields that are as productive as the world's best reserves in places like Iraq or Saudi Arabia. Nevertheless, it is indisputable that there are vast reserves of fuels left.

Figures from BP in Figure 2.1 show that estimated oil reserves have risen substantially in the last two decades.

▼ Figure 2.1 BP's figures for proved reserves of oil

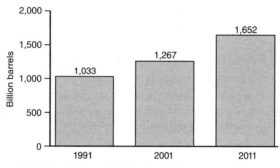

Source: *BP Statistical Review of World Energy 2012*

Although some people criticize the optimism of these figures, BP suggests that without any new fields the world has about 46 years of oil left at current rates of consumption. If we include the so-called 'unconventional' sources of oil from tar sand deposits and other sources, the reserves will probably last at least 100 years.

The same picture is true for gas (see Figure 2.2). The world has supplies to last at least half a century, even

▼ Figure 2.2 BP's figures for proved reserves of natural gas

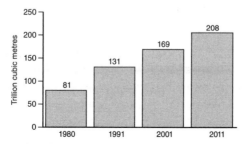

Source: *BP Statistical Review of World Energy 2012*

before considering the impact of hydraulic fracturing, or 'fracking', of shale rocks. 'Fracking' is a technique that increases available global reserves, perhaps by a factor of ten or even a hundred.

It is not worth even giving the figures for coal. Much of the world is sitting on coal that is cheap to mine and ship to remote power stations. Additionally, coal can be converted to gas and to oil. The crucial point is this: we are not going to run out of fossil fuels at any time in the near future.

At the beginning of this chapter, we said that sustainability requires us to meet two conditions – first, not depleting resources excessively and, second, not polluting or degrading the planet. In the case of fossil fuels, we have at least a century of reserves at current levels of use. In reality, we probably have more than this because of future oil and gas discoveries and better methods of extracting hydrocarbons from deep rock or in other difficult conditions. Yes, global energy needs will go up – perhaps by a third or more over the next 20 years – but fossil fuel use is not unsustainable in terms of the first condition we laid down. See Figure 2.3.

On the other hand, fuel use breaks the second condition, and does so in a spectacular way. The principal product of fuel combustion is CO_2 and our current levels of hydrocarbon consumption are responsible for about two-thirds of the rise in atmospheric levels of carbon dioxide. (Cutting down forests is the next most important source of greenhouse gases originating from human activities.) If BP's forecasts are correct, and assuming we are not able to capture carbon dioxide at source, greenhouse gas emissions will rise sharply in the years to 2030, and

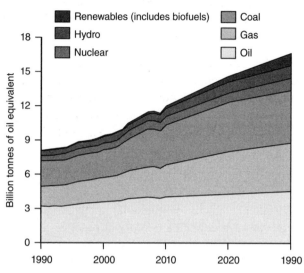

▼ Figure 2.3 BP energy use forecasts

Source: *BP Energy Outlook 2030*

advances in renewable energy will not significantly dent the use of fossil fuels.

▶ Minerals availability

When we burn a tonne of oil, we are turning a set of complex hydrocarbon molecules into carbon dioxide. Millions of years ago this oil was formed from decaying algae. The algae had employed photosynthesis to take the sun's energy and make complex and oily hydrocarbon molecules out of water and carbon dioxide. Burning oil neither creates nor destroys the carbon in these molecules; it moves them from safe storage in underground oil to the atmosphere, where they help retain heat by trapping the sun's energy. Modern society's problem is that we cannot easily collect the CO_2 in the air and return it safely underground. Capturing carbon dioxide in the air, where it still forms less than 0.05 per cent of the constituent gases, is difficult and very expensive.

Minerals are different. In the case of metals, we extract an ore from the earth, much like oil. Perhaps the metal forms 1 per cent of the material mined, or an even smaller fraction in the case of valuable substances like gold or platinum. When we refine the ore, we end up with nearly 100 per cent metal. In general, this metal is then used directly to make something useful. Copper becomes wire or an electrode. Aluminium is used for such things as packaging and car parts.

The crucial thing is that the metal usually remains in a highly concentrated form. At the end of the life of the building, for example, steel can be extracted, smelted and reused. Unlike CO_2, which cannot easily be recaptured, metals can generally be recycled indefinitely. It is usually far easier and cheaper to reuse a tonne of metal than it is to extract it from the original ore. Rarer metals have generally increased in price several-fold over the last couple of decades, sometimes making it far more commercially attractive to reuse them rather than search for new ores.

There are two central points to be aware of:

1 Metals never disappear and do not generally change their state in the way that carbon molecules do as they move from a hydrocarbon to carbon dioxide. Some metals are used in what is known as a 'dissipative' form – essentially meaning that they are spread very

thinly – such as when zinc is used to galvanize iron. It can be very difficult to get the original metal back, and some small proportion will always be lost, but most metals, once refined from ore, will stay with us for ever.

2 We probably don't require an indefinitely increasing amount of metal as we get more prosperous. The optimistic case – which I confess to believing – is that we need a good stock that meets our needs for buildings, cars, domestic appliances and consumer electronics. Once we have refined that weight of metal, we will simply reuse it carefully because it is easier to do this than refine expensive ores. In the chapter on iron and steel, I will suggest that the world needs a stock of about 10 tonnes per person and that amount is available from known reserves of ore of an adequate grade.

It is important to recognize that there are losses of metals and other minerals when we reuse them – not all metal will be eventually recycled. Perhaps a maximum of 90 per cent will be, even if the world gets extremely good at recycling metals. So there will be a need to continue to mine ores and then refine them at least until the world reaches peak population in around 2050. After that point, as population falls, the total required stock of metals may actually decline.

There is little argument with the view that the common metals, such as iron, aluminium and magnesium, are available in large enough quantities for our use to be sustainable for ever. In the case of iron, the numbers are given in Chapter 7. For some of the rarer metals, this optimism is a little less justified. Table 2.1 gives estimates of world production, 'reserves' and 'resources' produced by the US Geological Service (USGS) as at 2011.

▼ Table 2.1 Estimates of world production, 'reserves' and 'resources'

	Copper	Zinc	Lead
Annual production (million tonnes)	16.1	12.4	4.5
'Reserves' (million tonnes)	690	250	85
'Resources' (million tonnes)	3,000	1,900	1,500

Source: US Geological Service (USGS), 2011

This is how USGS defines the two key terms – 'reserves' and 'resources':

> **Reserves:** *'Mining companies' supply of an economically extractable mineral commodity.'*
>
> **Resources:** *'A concentration of naturally occurring solid, liquid, or gaseous material in or on the Earth's crust in such form and amount that economic extraction of a commodity from the concentration is currently or potentially feasible.'*

What does the table show? In the case of copper, current annual production will exhaust current mining company 'reserves' in about 40 years, but nearer 20 in the case of zinc and lead. But 'resources' are enough to cover current levels of consumption for a century or more.

The stock of copper in rich countries is currently about 200 kilograms per head. If all the population of 2050 had this level of the metal available to them, the world would have to have a stock of 2.0 billion tonnes, using up most of the world's 'resources', even assuming no losses in the recycling process. Therefore there is a plausible argument that our need for copper is unsustainable.

This might be too simple a conclusion. The scarcity of copper is already apparent, as its price in mid-2012 was four times the level of the year 2000. Users have responded by reducing usage. Almost all metals have reasonably effective substitutes. In the case of some of the key uses of copper:

❯ plumbing needs can be replaced by plastics and

❯ electrical circuitry will be switched to glass fibre or to aluminium.

In general, the relative scarcity of a mineral will produce a rise in prices, the substitution of other chemical elements and an intensified search for new products that reduce the need for the increasingly rare or pricey commodity:

❯ Well-functioning markets will tend to ensure that the recycling loops that return the metal to refiners for immediate reuse are active and efficient.

❯ There are few circumstances when one material cannot be effectively substituted for another. Substitution may be expensive and inconvenient but generally we have a range of choices: if the supply of one mineral becomes scarce, another can take its place.

There are exceptions to this second generalization. For example, many modern wind turbines employ permanent magnets composed partly of a substance called neodymium. This 'rare earth metal' makes the magnetic field much stronger than it otherwise would be. There currently isn't any substitute with the same valuable characteristics.

Will we ever run short of neodymium? Many people assume, because of their name, that the rare earths are scarce, meaning that future wind turbines will not contain neodymium and will be less effective as a result. This is incorrect: rare earths are actually substantially more abundant than metals such as tin or lead. Even the rarest is 200 times more common than gold. There are few places where they are mined, and almost all today's extraction is carried out in China, but there will never be an absolute shortage.

The US Geological Service estimates that world reserves of rare earths are about 110 million tonnes. Annual use is little more than 130,000 tonnes, meaning that currently available stock will suffice for many hundreds of years at current rates. And even rapid continued growth of wind power, with consequently increasing use of rare earths, would not alter the picture much. One US government study suggested that if turbines provided one-fifth of all electricity in 2030, and 20 per cent of these employed rare earth magnets, it would add only 320 tonnes a year to the demand for neodymium.

Other than metals, the main minerals that we need are quarried rocks for building materials. By weight, the most important is limestone, the key ingredient of cement. Each year the world produces about 3.5 billion tonnes of cement, more than half of which is made in China as the country races to build factories, roads and homes. As it matures economically, China will probably reduce its yearly use of the commodity. Figure 2.4 shows that, as GNP per head increases, cement use rises sharply – China is in this phase now – but then falls.

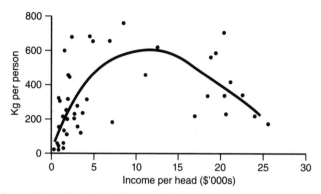

▼ Figure 2.4 Cement consumption per capita versus GNP per person

Source: Pierre-Claude Aïtcin, 'Cements of Yesterday and Today: Concrete of Tomorrow', *Cement and Concrete Research* (2000); reproduced in *Sustainable Materials: With Both Eyes Open* (UIT Cambridge Ltd, 2012)

As we would expect, countries like Britain and the US are seeing cement production fall a little each year.

The world's yearly cement production requires about 10 billion tonnes of limestone and other raw materials. This sounds a huge amount, but it corresponds to only about 4 cubic kilometres of rock, an infinitesimally small fraction of the total amount available to us. In this sense, cement use is fully sustainable. Unfortunately, to make cement the limestone has to be heated to a very high temperature (using fossil fuels) until the chemical bonds in the rock break down, driving off CO_2. So cement manufacture is an important source of global warming gases. The industry is responsible for about 5 per cent of world emissions. Since most of this total arises from the chemical reaction that breaks down limestone into calcium oxide and carbon dioxide, energy efficiency measures will have relatively little benefit. The only answer, described in Chapter 9, is to capture and store the CO_2 arising from the heating of the limestone, the process known as carbon capture and storage (CCS).

Unlike metals, cement cannot be cost-effectively reused. The world will have to continue making it, or find a realistic alternative for use in building. Exciting options do exist, some of which actually result in a net reduction in CO_2, but progress is slow in commercializing these substitutes. A world seeking to reduce emissions must invest heavily in the research and testing necessary to prove that these alternatives are as strong and durable as cement.

▶ Biomass

Biomass gives us the third group of materials we employ to give us a comfortable life. Biomass is anything directly or indirectly created as a result of the sun's heat and light including food, natural fibres and wood. It grows on land and, to a much lesser extent, in the seas. Some of the most acute sustainability challenges arise from the pressures that humankind is putting on the capacity of the Earth's top few metres of soil to give us what we need.

Food

This is such an important topic that Chapter 6 is devoted to exploring the main sustainability issues in food production. At this stage, all we need to say is that the world currently produces food crops equal to about

5,000 calories of food energy a head per day, about twice as much as required for humankind's nutrition. This might suggest we have enough calories for the larger population in 2050.

Nevertheless we have two main problems:

1 A large fraction of all food is eaten by meat animals, reducing the supply directly available to people.

2 An increasing fraction of agricultural output is being diverted to other uses, such as the corn used to manufacture ethanol, a substitute for petrol.

Agriculture also needs artificial fertilizer to create high yields. Fertilizer is made from natural gas, increasingly widely available around the world. Today's highly productive fields benefit from two other main fertilizers – phosphorus and potassium. Phosphorus is available in a relatively small number of rich deposits, some of which are in places such as the Western Sahara, a region of uncertain political stability. Current patterns of phosphorus use are possibly unsustainable but, as with metals, we can develop more effective ways of ensuring that the available supplies are recycled, principally at sewage plants. Potassium is abundant.

Productive farming also often necessitates irrigation. Water use is clearly running down natural aquifers in many places around the world and exhausting the flow of rivers. Although climate change will increase the total amount of world precipitation, drier areas will generally get less than they do at the moment, creating scarcities of fresh water in some of the world's most important crop-growing areas.

Fibre and wood

Cotton production, the most important non-food use of arable land, uses only about 4 per cent of agricultural cropland so it is not a major competitor to food production. Furthermore, Chapter 8 on clothing suggests that the use of artificial fibres has grown at the expense of natural material. A larger and richer population will not necessarily mean an increase in the acreage devoted to cotton. Wool production from sheep and goats does not usually cause competition for scarce arable land.

Woodland is a much more difficult issue. Increasing demand for food increases the economic incentive to cut down forests to create new fields to grow crops. Forests absorb carbon dioxide from the atmosphere through photosynthesis, so the loss of woodland is a major cause of rising greenhouse gas concentrations. Retaining forest cover, and preferably increasing it, is a prime aim of global policy.

This quick analysis shows that we probably have enough fuel, minerals and biomass to meet the needs of a larger and richer world population in 2050 if sensible policies to conserve resources are put in place. The world should be able to provide for itself and will not suffer unduly from the depletion of resources. We cannot offer the same optimistic conclusion about the second form of sustainability: the need to avoid dangerous pollution. Our use of fossil fuels, the use of limestone kilns to make cement, and the intensification of agriculture are all threatening the future.

Will economic growth make our problems worse? An unconventional view

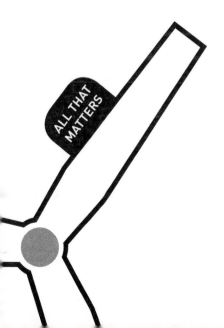

Will economic growth make our problems worse? An unconventional view

30

The cautiously cheerful conclusion about resource availability in the last chapter is based on my assumption that, although developing countries are using more resources, the rich world's consumption will not grow very much. This is in sharp contrast to the conventional view that economic growth inevitably draws in large volumes of extra physical resources, such as energy and metal ores, even in the richest economies. Most sustainability specialists say that 'continued economic growth is impossible on a finite planet'. However, if economic growth greatly improves the efficiency with which we use materials, or if eventually we reach satiation of our needs for physical goods, then the conventional wisdom may be wrong. Growth may be compatible with a sustainable society.

Historical evidence does not immediately support my optimism. Since the beginning of the Industrial Revolution improvements in prosperity have largely been delivered to us through increases in the availability of minerals, fuels and biological materials such as food and wood. Measurements of these flows are difficult and we can have limited confidence in today's numbers, let alone those from centuries ago. But we can make estimates. Austrian researchers led by Fridolin Krausmann are at the forefront of measuring trends in resource use. Their conclusion is that the average person in Europe consumed about 8–15 kilograms of raw materials per day in 1800, equivalent to 3–5.5 tonnes a year. This provided food, clothing, shelter and energy. The typical individual had access to about 40 kilowatt hours of energy per day, mostly provided in the form of wood fuel but also through the labour of animals or windmills for grinding flour.

In European countries today the figure is about 16 tonnes a year, four times as much as in 1800. As the developing world progresses, we can assume that per capita consumption in poorer countries will eventually rise from today's figure of about 4 tonnes (similar to Europe before industrialization) to a much higher figure. (For a comparison of the material per capita consumption of some of the world's geopolitical areas, see Figure 3.1.)

The crucial question is whether newly industrializing countries copy Japan, which has very successfully held down its use of raw materials, or the USA, a far more extravagant user of planetary materials. Japan's use of material resources had already peaked by 1990 and fell by a further 29 per cent between that date and 2004. Japan has shown that an economy can grow – albeit very slowly – and reduce the amount of material it uses. The pattern is very different in the US – one group of researchers writes of the continuing connection between economic growth and materials use:

> At the turn of the twenty-first century, each additional dollar of GDP still requires roughly an additional 0.2 kilograms of renewable biomass, 0.7 kilograms of mineral and fossil materials and 11 MJ (about 4 kWh) of primary energy.

Krausmann concludes that across the world as a whole:

> The shift towards a post-industrial or service economy which has been observed since the 1970s

Will economic growth make our problems worse? An unconventional view

32

▼ Figure 3.1 Domestic material consumption per capita per year

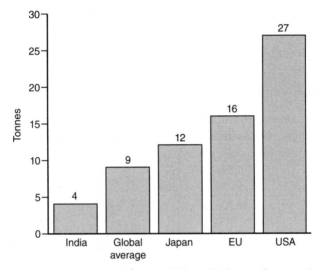

Source: *Resource Use in Austria* (Austrian Ministry of Agriculture, Forestry and Environment, 2011)

> *has not resulted in abolishing the energy and material intensive structural characteristics of the industrial regime or in a significant decline in natural resource use in industrial economies.*

The pessimistic case, made most compellingly by Tim Jackson in *Prosperity without Growth*, is that the developed world's use of resources will continue to grow a little less fast than GDP, a relationship known as 'relative decoupling'. For growth to be fully sustainable, as he correctly points out, we need to have 'absolute decoupling', meaning that the use of resources is flat or declining even when the economy is growing. And the European evidence suggests that we are seeing relative, but not absolute, decoupling. At the peak of the economic boom in 2007, material consumption per head of population was up about 5 per cent since 2000. Since this figure excludes some of the resources used in imports, the actual rise in resource use may well have been greater.

The UK may be more like Japan and less like the rest of Europe. Recent evidence suggests the growth of the economy has been decoupled from the use of materials. The things we buy and sell in Britain are increasingly 'dematerialized'. Our additional purchases today are more likely to be online music, some adult education or perhaps a trip to an art gallery to see a new exhibition. Compared to the past, our spare money is increasingly likely to be spent on services rather than on increased amounts of food, more cars or bigger houses.

Although the picture is made less clear by the recession of the last few years, which may be giving an unduly optimistic sense of progress, the weight of goods flowing through the economy has fallen sharply. What statisticians call the 'Total Material Requirement' of the British economy, marrying domestic extraction of resources with the volumes embodied in imports into the UK, was no higher in 2009 than it was in 1970. The fall since the peak around 2005 has almost certainly continued.

We can split the resources used into the three constituent parts – biomass, fossil fuels and minerals. About 96 million tonnes of UK biomass in the form of

Will economic growth make our problems worse? An unconventional view

34

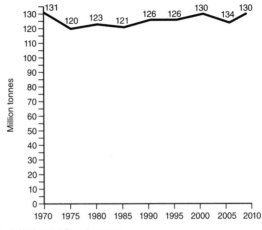

▼ Figure 3.2 Biomass consumption in the UK

Source: ONS Material Flow Accounts

agricultural products, timber and fish was harvested in 1970, complemented by 38 million tonnes of imports less 3 million tonnes of exports. Total usage was about 131 million tonnes (see Figure 3.2). By 2009 this figure was down slightly to 130 million tonnes, even though the UK's population had risen by more than 10 per cent in the period from 1970 and inflation-adjusted GDP had risen by a factor of nearly 2.5.

The decline is more obvious in the case of fossil fuels. The total weight of oil, gas and coal used by the economy declined from 266 million tonnes in 1970 to 208 million tonnes in 2009 (Figure 3.3).

The pattern is similar with minerals, comprised of ores, building materials, clay and other industrial materials mined from the top few kilometres of rock (Figure 3.4).

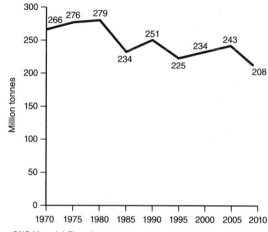

▼ Figure 3.3 Fossil fuel use in the UK

Source: ONS Material Flow Accounts

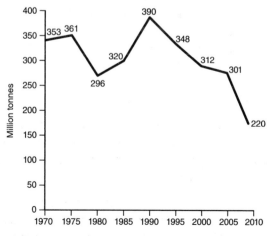

▼ Figure 3.4 Mineral use in the UK

Source: ONS Material Flow Accounts

Will economic growth make our problems worse? An unconventional view

36

Add these three categories of input to the economy together and the total is a reasonable estimate of the ecological strain put on the UK from the extraction of resources. This figure has declined sharply since the mid-1990s.

This is not the complete picture. Although the weight of goods imported from abroad is measured in these accounts, the inputs used to make these things in a foreign country, much of which have remained there as waste, is not included. Statisticians provide a second measure that includes an estimate of the weight of resources used abroad to make the items that were sent to the UK. As Britain has de-industrialized over the last few decades, an increasing fraction of the country's total impact on the global environment has occurred overseas. But even taking into account the increase in foreign-made goods, and the relatively generous amounts of materials and energy used to make these things, the total weight of inputs into the economy has fallen – but only just – since 1970 (see Figure 3.5).

We should not overestimate the reliability of these figures. The estimates of the total resources used in foreign countries to make goods for export are imprecise. Nevertheless, the pattern is reasonably clear: economic growth does not have to be the disaster for the environment that we once thought. As an economy grows it usually becomes more efficient, using fewer resources to create a growing volume of goods or services. Technological change, which is encouraged by increasing GDP, tends to reduce the weight of the things we consume. This is not surprising: heavy things contain

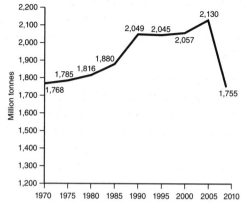

▼ Figure 3.5 Total material use in the UK

Source: ONS Material Flow Accounts

more materials and those materials are often expensive. All other things being equal, a business using fewer physical resources to manufacture its goods will prosper in competition against a less parsimonious supplier.

The value of thrift in the use of raw materials becomes more apparent in a world of increasing prices for metals and other scarce resources. As Figure 3.6 shows, the market price of important commodities like copper has often risen sharply in recent years as world demand has increased and the mining companies use lower-grade ores.

We are also tending, slowly but apparently inexorably, to reach the limits of our need for physical goods. In advanced economies most adults now have access to a car, own most of the kitchen appliances that they need, have little interest in accumulating more furniture and,

ALL THAT MATTERS: SUSTAINABILITY

Will economic growth make our problems worse? An unconventional view

38

▼ Figure 3.6 Copper prices since 1982 in $ per tonne

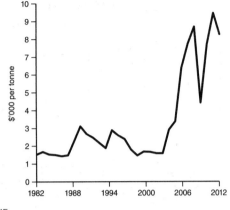

Source: IMF

difficult though this is to believe, are eating less than they used to. Even the consumption of meat – usually marked out as being closely tied to the prosperity of a country – has been tending to fall in the UK.

In other measures of the demands Britons are placing on the ecology of the planet, they are travelling less and using less water. The one exception may be clothes, where the appetites of British people may not yet be sated. But, even in this case, the last few years have seen a decline in the weight of clothing bought.

If total amount of material entering the economy is tending to fall, then eventually we should expect to see declining volumes of waste leaving it at the end of useful life. An economy is like a living organism, taking in useful nutrients that eventually are expelled as waste. And, at

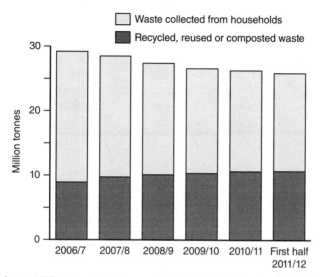

▼ Figure 3.7 English local authority collected waste

Source: UK Department of Environment, Food and Rural Affairs. Local Authority Collected Waste for England

least in the UK, waste volumes are indeed tending to fall. (In addition, and quite separately, more of what is disposed of by households is then recycled or reused rather than being sent to landfill.) Figure 3.1 shows the figures for all waste collected from households. The figure for businesses shows a similar decline.

The UK may well not represent a typical picture. But its unusual characteristics – such as the openness to digital technology and the increasing interest of its population in spending on experiences rather than buying new physical goods – might well represent the direction of global society. Growing technological efficiency, better

Will economic growth make our problems worse? An unconventional view

40

reuse and recycling, greater durability, and satiation of the demand for buildings and machines, including cars and domestic appliances, could mean that economic growth can continue without enhancing the strains on planetary ecology.

When I wrote about the dematerialization of the UK's consumption in an article entitled 'Peak Stuff', several commentators complained that I was reading too much into what might just be a short-term trend that will rapidly reverse when the economy starts growing quickly again. But the confluence of forces pushing societies to use smaller quantities of material resources is powerful. Examples include:

▶ rising prices of fuels, foods and metal ores

▶ increasing pressure to reduce waste because of concerns over the availability of landfill sites

▶ regulatory drives to reduce the production of climate-changing gases

▶ social trends – populations are ageing almost everywhere and older people seem to consume fewer material resources

▶ the increasing availability of digital equivalents – in the world of media, for example, online newspapers, ebooks, games on mobile phones, and online video are replacing physical media.

In the case of Japan, the drive towards dematerialization started after the oil shortages of the mid-1970s. The country realized that its almost total reliance on imported energy made the economy highly vulnerable to war and political disputes elsewhere in the world. Politicians continue to give high priority to reducing the weight of inputs into the economy and are targeting a 60 per cent increase in Japan's 'resource productivity', the amount of GNP generated per tonne of material entering the economy. This target – and similar objectives to reduce waste and increase recycling – are challenging targets that politicians and administrators are struggling to deliver. Researcher Fridolin Krausmann says that, while Japan shows what is possible, the country also shows the difficulty of getting global agreement to take the required actions:

> The case of Japan shows very clearly that while severe (and successful) efforts have been made to reduce materials use and increase resource efficiency, laying a path towards sustainable levels of resource use still remains a major political challenge.

The last two chapters have focused on the first of the sustainability questions: Do we have enough materials to give everybody in 2050 the standard of living of a reasonably prosperous person today? The next section looks at the second question: Are we able to avoid dangerous disruption of the planetary ecology? As you will find, I think we should be much more worried about the answer to this query.

4

Planetary boundaries

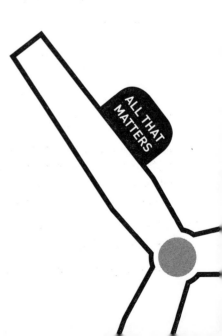

Johan Rockström leads a team of researchers at the Stockholm Resilience Centre. As its name implies, this scientific institute looks for ways of improving the planet's ability to withstand the unprecedentedly rapid changes in Earth's ecological systems happening at the moment. His research group originated the idea of 'planetary boundaries', a set of ecological limits within which humankind must stay.

In an influential but contentious 2009 publication, Rockström and colleagues listed nine interrelated boundaries. Crossing any of these borders, they said, moved humanity from a relatively safe position to potential catastrophe. Staying within what they called 'the safe operating space for humanity' minimizes the chance of triggering abrupt and dangerous environmental change. Whatever you think of the measures that the Swedish team proposes, the idea of planetary boundaries provides a powerful and thought-provoking way of looking at the second of the sustainability challenges: the need to avoid pollution and disruptive changes to global ecology.

Rockström's team suggests nine different boundaries. As is probably obvious when you look at the list below, many are directly related. For example, greenhouse gas levels affect ocean acidity and, through climate change, the level of biodiversity. Nitrogen levels also affect biodiversity, while water abstraction is closely connected to the amount of agricultural land. The nine boundaries are:

1 Biodiversity levels and, in particular, the rate of extinction of species

2 Greenhouse gas levels in the atmosphere

3 The amount of nitrogen and other artificial fertilizers applied to fields

4 The percentage of the world's surface area used for growing crops

5 The amount of fresh water abstracted from rivers for irrigation and human use

6 Levels of toxic chemicals in the atmosphere, in water and on land

7 Atmospheric pollution caused by soot, very fine particles and trace gases such as sulphur dioxide

8 Changes in the acidity of the oceans

9 The density of ozone in the stratosphere.

▶ Biodiversity

Humankind's actions are causing the rapid loss of species of animals, insects and plants. Many have called this 'the sixth great extinction', comparing today's changes to previous events such as the end of the dinosaurs and the ensuing rise of mammals. Rockström's group contends that the current rate of species loss is perhaps 100 to 1,000 times the background rate over the planet's history. They say that extinctions need to be reduced by a factor of between ten and 100 to keep humanity within safe limits, although they provide very little evidence as to why this is the right number.

Until recently, the greatest loss of biodiversity occurred on ocean islands where, for example, the introduction of rats has often brought about the rapid destruction of other small animals. However, as the influence of humankind has grown, a larger and larger fraction of species extinction has been on large landmasses. 'Biodiversity is now broadly at risk,' the scientists wrote. Many of the largest and most photogenic of animals, such as orang-utans, tigers and gorillas, have diminished sharply in number and will probably disappear completely within a half-century.

The risk voiced by the researchers is that today's rate of extinctions is likely to prompt major changes in ecosystems, causing irreversible damage. The destruction of the Newfoundland cod referred to in Chapter 1 prompted a cascade of other effects, causing what may be permanent harm to an area's capacity to provide fish and other services for humanity. Are other events similar to this going to cause other long-lasting changes to important ecosystems? We cannot be certain, but the example of the last five major extinctions clearly shows the possibility of fast changes in vegetation types occurring alongside species loss. On the other hand, while the loss of important species such as polar bears may be an ethical disgrace, it is unclear whether it has substantial impact on global (human) sustainability.

Rockström's attention focuses on the rate of complete extinction of species. This seems to me to be questionable as a proxy for a loss of sustainability. At the scale of a few hundreds of hectares – for example, a large farm – what seems to prompt the unhealthy ecological changes that he worries about is a rapid reduction in the diversity of species in that specific area. A monoculture, perhaps growing only one or two crops using large open fields, seems to be much less resistant to pests and have lower overall agricultural yields than areas with many crops, sometimes intercropped in the same field, and patches of varied vegetation and diverse habitats. In other words, if biodiversity matters to sustainability, it may be at the very small scale rather than as an absolute amount across the entire globe.

▶ Greenhouse gas levels in the atmosphere

Much of this book is about the dangers of climate change and what we need to do to arrest the growth of CO_2 in the atmosphere and there is little need for comment here. But one additional thing worth saying is that Rockström's team, including the eminent climate scientists James Hansen and Hans Joachim Schellnhuber, indicates that the boundary should be set at 350 parts per million (ppm) of CO_2, a long way below the current level of about 400 ppm. The scientists believe that global temperatures must be restricted to a rise of less than 2 degrees centigrade above the levels of 200 years ago to avoid tipping Earth into increasingly severe impacts. They contend that the temperature record over the last hundreds of millions of years demonstrates that a CO_2 concentration of more than 350 ppm is incompatible with this aim.

Many other climate specialists have contested the logic behind this very tight proposed boundary, which requires humankind to reduce net greenhouse gas emissions to below zero within the next few years. Most still think that the figure should be about 450 ppm, a number broadly compatible with Myles Allen's suggestion, covered in Chapter 9, that the world cannot cumulatively burn more than a total of a trillion tonnes of carbon fuels.

▶ The amount of nitrogen added to the environment

In the atmosphere, individual atoms of nitrogen are tightly bound to a second atom, forming a very strong coupling. Atmospheric nitrogen is of no use to most plants, which need this chemical element in order to grow, because the link between the two atoms cannot be broken. But in chemical fertilizers single atoms of nitrogen in highly reactive molecules can be taken up productively by vegetation as a nutrient. Today's high yields from our crops are partly due to the beneficial effect of nitrogen and phosphorus fertilizers on plant growth. Although some plants, such as beans, use microbes on their roots to break down twinned nitrogen atoms in the atmosphere, the world's main crops are dependent on the nitrogen provided in artificial fertilizer.

Some – perhaps half – of the nitrogen in fertilizers is not taken up by plants and runs off into watercourses. The world's rivers and seas are chronically short of nutrients and the addition of a powerful fertilizer causes rapid growth of some organisms, such as algae, which eventually starve the water of oxygen, creating what are usually known as 'dead zones'. Other detrimental effects of nitrogen application to crops include a significant loss of biodiversity as more vigorous plants crowd out less robust varieties.

Rockström and his team suggest tentatively that the world's creation of so-called 'reactive' nitrogen, which also arises from fossil fuel burning, needs to be cut to no more than a quarter of today's levels. They call this 'a first guess only'. Cutting the amount of reactive nitrogen in the environment is an important objective but may be costly to achieve. World food supplies would fall sharply if nitrogen applications to fields were reduced.

Similarly, forest soils have been enhanced by nitrogen compounds in the rain helping trees grow faster, increasing absorption of atmospheric CO_2. Balancing the need to improve total food yields and maximize biomass growth against the less urgent need to decrease the amount of nitrogen on land and in water is a difficult problem for the world to address.

▶ Percentage of land area used for growing crops

The first three of Rockström's boundaries are set lower than today's levels. In their view, we are already beyond the safe limit. For the remaining six, the researchers say the world is still within the boundary. The team suggests that the appropriate limit for cropped agricultural land is about 15 per cent of total land area, about 3 per cent higher than today's extent. The most important reason for setting a limit is that the extension of cropland comes almost exclusively from the expansion of agriculture into tropical forests. Other areas are usually too dry to grow crops. One recent scientific paper by Andrew Balmford's team comments that:

> Agriculture is by far the leading cause of deforestation in the tropics and has already replaced around 70 per cent of the world's grasslands, 50 per cent of savannahs and 45 per cent of temperate deciduous forest.

We need forests to soak up carbon dioxide from the atmosphere and to maintain a relatively stable pattern of worldwide rainfall. Many studies show that the loss of major areas of tropical forests such as the Amazon will result in painfully significant changes in rainfall distribution: an area getting plenty of water today may become parched in a world of fewer forests while other landscapes will be flooded more often.

The current rate of forest loss is about 6–7 million hectares a year, slightly less than the size of Scotland and twice the size of the Netherlands. This is about 1/600th of the total extent of forest land in the world. At the current rate of conversion to cropland, Rockström's proposed boundary will be breached about 50 years from today. Will the need for arable land, and thus the rate of conversion of forest, increase from today's levels and therefore reduce the length of time before the boundary is reached? Put simply, it mostly depends on diet. If meat consumption keeps expanding, the limit will probably be reached very much sooner. Chapter 6 has more details.

▶ Fresh water abstracted from rivers

Rockström and his team estimate the world uses about 4,000 cubic kilometres of fresh water a year for irrigation from run-off into rivers and ditches. This is close to their proposed limit of 4,000 to 6,000 cubic kilometres. Breaching the boundary will result in the destruction of many habitats as rivers become exhausted. Already, Rockström says, 25 per cent of rivers dry up before reaching the sea. Many major river systems, such as the Murray–Darling in Australia or the Colorado in the US, are at the limits of what can be extracted safely. The collapse of lakes, such as the Aral Sea, as a result of excess abstraction by agriculture threatens regional water shortages and destructive changes in local ecologies brought about by rises in salinity and increased toxicity of soils.

Most projections see substantial increases in the need for irrigation over the next few decades as food demand increases. (A very large amount – perhaps as much as 80 per cent – of all abstractions from river water are used for agriculture.) Without more efficient use of water – encouraged by irrigation channels with less evaporation and leakage, far better targeting of the water onto the plant, and the development of drought-resistant crops – the world will almost certainly run up against the limited supplies from rivers, and will have to rely instead on rapidly depleting aquifers.

▶ Levels of toxic chemicals

The researchers do not set a precise boundary for chemicals that are dangerous to human health. However, of the 80,000 or so chemicals currently being made by industry, 200 are known to be toxic to the nerves or brains of human beings. Some 'Persistent Organic Pollutants' (POPs) are now widely dispersed across the globe, and do not exist only in localized concentrations near factories or in areas where the chemical is used. Some of these pollutants may now be causing damage to planetary ecology. Nevertheless, Rockström's team is uncertain of how to set a boundary or what that limit might be.

▶ Atmospheric pollution

High levels of tiny particles in the atmosphere produce substantial effects on climate across large areas. These effects are particularly pronounced in Asia where pollutants from fossil fuel burning and from open wood fires affect the location and intensity of monsoon rains. In addition, these particles cause respiratory and heart disease in humans.

Some types of atmospheric pollution may actually work to counteract the effects of greenhouse gases in the atmosphere, tending to reduce the amount of trapped solar energy. This, and the many uncertainties as to the limits that should be set, means that Rockström does not set a boundary for air pollution.

▶ Ocean acidity

When sea water dissolves CO_2 – which is an important natural way of reducing greenhouse gases in the atmosphere – the water moves to a less alkaline state. As carbon dioxide levels rise in the atmosphere, more of the gas ends up dissolved in sea water, meaning that the world's oceans will gradually tend to become more acid. The speed of this change is disturbingly rapid: ocean acidity is changing at least 100 times faster than at any point in the last 20 million years. Ecological systems affected by this change will almost certainly not have time to adapt.

Many marine organisms such as molluscs and corals need the water in which they live to be alkaline to help protect and grow their calcium shells. Acid water dissolves these shells, or reduces their ability to develop to full size. Ocean acidification may be a very serious problem because organisms using carbonate form a vital part of marine ecosystems. 'The consequences and impacts of this are highly uncertain,' write Rockström and colleagues.

In addition to the impacts on small ocean creatures, increased acidity adversely affects corals, reducing their ability to grow. Corals provide a habitat for huge biodiversity and shelter fish that are important sources of food for humankind. It is conceivable that waters that have warmed as a result of climate change will improve the average growth rate of corals around the world, but the destructive impact of increased acidity is likely to hugely outweigh this positive effect. Already many coral reefs in places such as the Caribbean and the Great Barrier Reef are showing the impact.

The planetary boundaries research team does not propose to set a boundary for ocean acidity itself but suggests a lower limit for the density of the key chemical in sea water that controls the ability of sea creatures to grow their shells. The quantity of this substance – aragonite – in sea water should not fall below 80 per cent of the pre-industrial level. It is not far from this limit today.

▶ Density of ozone in the stratosphere

Ozone high up in the stratosphere has a crucial function protecting living things from destructive ultraviolet (UV) radiation. As is well known, UV causes skin cancer in human beings, but it also affects small marine organisms and crops. A class of chemicals developed to safely operate refrigerators and air conditioners the best part of a century ago reduced the amount of ozone in the stratosphere. These chemicals, usually referred to as CFCs, were made in quite small quantities but had an enormous impact on the chemistry of the higher levels of the atmosphere, inducing completely unexpected changes. More than any other single example, ozone depletion shows that human actions can rapidly tip the environment into a very different, and profoundly dangerous, state.

International agreements in the late 1980s (usually called the Montreal Protocols) were successful in forcing manufacturers to stop making CFCs and the dangerous destruction of the ozone layer has been partly reversed. Full restoration of ozone levels in the stratosphere around the polar regions will take several decades more. As Rockström's team says, 'Stratospheric ozone is a good example where concerted human effort and wise decision-making seem to have enabled us to stay within a planetary boundary.'

▶ The strengths of Rockström's approach

The planetary boundaries idea has clear value. It recognizes that Earth is a complex organism with many closely connected ecological systems, all of which need to be managed to ensure a relatively stable and liveable planetary environment. It also helpfully suggests that, although humankind has severely disrupted almost every single natural process, the consequences need not be disastrous. In effect, Rockström's team says that the planet probably possesses the capacity to adjust to quite large perturbations from pre-industrial norms. For a safe future, we do not have to return to the natural state prior to the Industrial Revolution.

There are nine proposed boundaries but two of them are utterly central. The first is greenhouse gas emissions. Climate change resulting from increases in carbon dioxide levels in the atmosphere is affecting biodiversity, water supply and ocean acidity. Agriculture is crucial to the land use and water boundaries as well as to the nitrogen limit. In fact, you could reduce the nine boundaries to two simple rules:

1 Stop burning fossil fuels as fast as we can.

2 Ensure that agriculture uses as little land, water and fertilizer as possible.

▶ The weaknesses

Critics of Rockström's approach say that the limits seem arbitrary and not scientifically robust. The boundaries have not received wide acceptance from scientists unconnected with the project.

The criticisms of the planetary boundaries idea can be divided into four main groups:

▶ *Earth systems are not as stable as the research group suggests.* It is not good enough to suggest that, as long as the planet remains within a series of arbitrary limits, it will be safe. Every move away from the pre-industrial position carries dangers because we often do not understand what tips ecologies from one state to another.

▶ *The limits proposed by the group are little more than guesses.* This point is acknowledged by Rockström and his colleagues, who carefully indicate the lack of certainty about the right limit for any of the variables. They clearly state that their paper only sets out a framework for considering what the planetary boundaries are rather than providing a definite limit.

▶ *Even the limit for the best-researched boundary – greenhouse gases – is wholly uncertain.* The Rockström paper proposes 350 ppm, substantially lower than present levels, but most climate scientists suggest a much higher number that the world will not reach for about 20 years. How useful, we may ask, is the notion of a 'boundary' when the value of that limit is impossible to define with any reasonable certainty?

▶ *The idea of limits allows humankind to be sloppy.* As William H. Schlesinger pointed out in a response to the Rockström paper, 'Setting boundaries is fine, but waiting to act until we approach these limits merely allows us to continue with our bad habits until it's too late to change them.'

5

Two central ideas

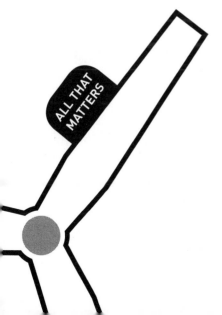

ALL THAT
MATTERS

In the first three chapters we focused on resource availability, before turning to pollution in Chapter 4. This chapter examines two crucial concepts that seek to help us reduce use of material and minimize the consequence of human activities on the planetary environment.

▶ 'The circular economy'

Put at its simplest, the twentieth century was the time when humankind viewed Earth's capital as a cheap and apparently limitless source of goods and services such as energy. It was the era of near-universal 'take-make-dispose', as one source puts it. Although the price of primary commodities sometimes varied sharply from year to year, the long-run trend was downwards for almost all natural resources, including foods and metals. For most of the last century, it looked as though the Earth's crust was a cornucopia, delivering unlimited amounts of natural resources. Not surprisingly, humankind engaged in an undignified scramble for the cheapest oil and easiest ores, ignoring the benefits of energy efficiency or of recycling.

It was not always like this. Before industrialization, most societies had little choice but to live within much tighter constraints. Extracting metal ores from the rocks beneath us, mining coal for fuel, cutting stone for new buildings or growing more than a subsistence amount of food on unfertilized soils was almost invariably difficult. As a result, far more emphasis was placed on keeping goods for longer, repairing and refurbishing as much as possible, and then eventually recycling the items for further use elsewhere. Virtually nothing was 'waste' in the sense we understand it now. Yes, we find medieval tips containing broken pottery, animal bones and the remains of other organic material, but metals, stone and even human excretions were rarely wasted. As we realize with increasing clarity, these parsimonious habits of the past will have to be resurrected to meet the circumstances of the twenty-first century.

In 2002 William McDonough and Michael Braungart published their book on the recycling of materials in an endless loop. Called *Cradle to Cradle*, it was printed on a paper made not from wood pulp but from plastic resins and other reusable materials. The authors stressed the symbolic importance of this feature of their book. They said:

> This material is not only waterproof, extremely durable, and (in many localities) recyclable by conventional means; it is also a prototype for the book as a 'technical nutrient', that is, a product that can be broken down and circulated indefinitely in industrial cycles, made and remade as 'paper' or other products.

For many, *Cradle to Cradle* was their first introduction to the idea of the 'circular economy', a vision of how humankind could re-engineer its production processes and industrial design to minimize the need to mine the Earth for new minerals, biomass and fuels. Instead of continuously making things that are used for a limited time and then thrown away without reuse, *Cradle to Cradle* envisaged a process of continuous cycling of finished products so that little will ever become waste. They stress the difference between true reusability, which enables something to circulate almost indefinitely, and what they call 'downcycling', the process of employing materials in products of lower and lower value. Their own book can be melted down into raw plastic and then used again and again, perhaps to create updated editions. The paper a newspaper is printed on is, by contrast, 'downcycled' since its constituent fibres have been shortened every time they have gone through a paper mill. After four or five uses, these fibres become unusable.

To an extent that may be unrecognized, biological materials such as paper, natural fibres and wood are much more of a challenge to sustainability than artificial products such as plastics. We can put most mineral products into infinite recycling loops but natural products are much more likely to degrade. Furthermore, materials originally grown in fields or woodlands add to the pressure on the world's land resources. As Chapter 6 will show, we will struggle to meet the world's needs for food and the more we can reduce the space needed to grow other raw materials, the better our food security will be.

McDonough and Braungart's focus is not simply on reducing the need for newly extracted raw materials but also on ensuring that the constituent elements in the world's stock of products represent no threat to human health. We do not want toxic chemicals circulating continuously as they are recycled from one product to another. As importantly, a circular economy reduces the need for inputs of energy. Recycling an aluminium can rather than making new metal reduces energy needs by 90 per cent or more. Two-thirds of European drinks cans are now recycled, meaning the soft drinks business has moved a long way towards neither depleting Earth's resources nor imposing new climate change stresses. This pattern needs to be replicated across the economy.

The first decade or so of the twenty-first century has demonstrated all too clearly that natural resources may well become far more expensive and difficult to extract. The long, slow decline of raw material prices seems to have ceased and sharp upward spikes have become common. Recycling of existing stocks of materials and the use of renewable forms of energy makes more and more financial sense. Economic pressures may already be making us adopt more sustainable practices without us really intending to move in that direction.

▶ Why don't we have a circular economy today?

The world economy is locked into production processes and industrial designs from the era when everything seemed to be getting cheaper. As a result, few manufacturing processes are currently set up to ensure the maximum reusability of raw materials and components at the end of the useful life of the product. Instead, companies engineer their products to achieve the lowest cost of initial manufacturing and have tended to ignore any benefits from creating a loop that collects and reuses the product if in so doing they increase the cost of initial manufacture even by a tiny amount.

For example, today's lowest-cost way of fixing a circuit board in a piece of electronic equipment is to use a tiny spot of glue. Using a small clip would be more costly initially but might significantly reduce the difficulty and expense of taking the device apart and reusing the components. In an economy with a full recycling infrastructure, manufacturers will be happy to use the clip. Until then, they will employ the older method.

A 2012 report from the Ellen MacArthur Foundation, a leading think tank promoting the circular economy, quotes the example of mobile phones as a good illustration of the issue:

> Typically weighing less than 150 grams, a mobile phone is packed with valuable materials such as gold, silver, and rare earth metals. Given today's low collection and recycling rates, nearly all of this material is lost. In Europe alone, for example, 160 million discarded but uncollected devices represent a material loss of up to USD 500 million annually.

To put mobile phone manufacture into a circular process many separate things need to happen, says the report. A system needs to be put in place that improves the lamentable collection rate for old phones; components need to be standardized so that, say, a camera used in one phone can be reused in another model, and full disassembly needs to be possible without damaging the product. Building a closed loop for the components of electronic devices will be not only directly beneficial for both customers and manufacturers but will also help industrial designers cope with rapid developments in technology.

This may sound like an eminently good idea but to companies such as Apple or Samsung the idea of standardizing on a product skeleton that is shared with their main competitors is highly problematic. To put this at its simplest, the world's needs for durability and recyclability are in conflict with our preference for unique design and rapid changes in the appearance of our products.

▶ Renting rather than owning

The other key focus of proponents of the circular economy is on the importance of a switch from customers owning goods to instead renting them from their manufacturers. The logic is this: if businesses generate their sales from selling goods, they have little interest in building highly durable and reliable products that can be fully recycled at the end of their life. But if their revenue comes from monthly fees for, say, washing machines, they will be motivated to build appliances that will last for a long time, require little maintenance and can be brought back and refurbished at the end of the contract life. Once again, the logic is sound but the commercial realities may be more complex – having been used to owning kitchen appliances, will householders switch to an arrangement that sees them paying a rental, whatever the advantages to manufacturer and consumer?

The previous chapter showed that, although the UK and Japan may have managed to begin the process of absolute decoupling of GNP from the use of material goods, the world as a whole is still increasing its use of physical resources. Rapid economic growth in poorer countries – a process that has pulled many hundreds of millions of people out of poverty – will continue to cause resource needs to rise and increase the threat to climate stability. Even if we have enough raw materials in the ground to give us what we need for 2050, it makes sense to drive as fast as we can towards circularity.

The second imperative is to measure the footprint of everything we do, and manage the impact downwards.

▶ Footprinting

The central idea of the circular economy is that materials such as metals should be used in durable, long-life goods that can be disassembled at the end of their life, reprocessed and then reused with minimal loss. The notion is particularly applicable to physical objects such as buildings, appliances, consumer electronics and cars. A circular economy principally applies to minerals, and not as much to the categories of biomass and fossil fuels.

For biomass and fossil fuels the core concept is of Earth's *biocapacity*, the amount of material it can provide to sustain human activities and dispose of the wastes that result. The best-known measure of the relationship between humankind's requirements and Earth's capacity to absorb our impact is called the 'ecological footprint'. This term was invented by Mathis Wackernagel, a Swiss scientist now working at the Global Footprint Network, and his PhD supervisor William Rees.

An ecological footprint is a numerical measure that tots up the amount of land needed to cater for human needs and is therefore expressed in hectares. Our requirements include areas for growing food, woodland for fuel, paper and timber, and space for roads and buildings. Of critical importance is the final and largest need for geographical area – the area of forest required to sequester the CO_2 produced from the combustion of fossil fuels.

Estimates produced by Wackernagel and his colleagues suggest that the ecological footprint of humankind exceeded the area of the world's land surface in about 1970. Since then, humanity has been in ecological deficit, demanding more space than the land area of Earth can provide. In recent years, the footprint of our activities has exceeded 1.5 times the world's land area. Projections suggest that this ratio will rise to more than 2 by 2030.

A quick look at Figure 5.1 shows where the principal problem lies. About 55 per cent of humankind's ecological footprint arises from the large areas of land required to absorb the CO_2 from fossil fuels back into living biomass. The world does not have that land, so the carbon dioxide is going into the air and oceans, causing climate change and changes to the sea's acidity. Simply put, say followers of this approach, the levels of today's

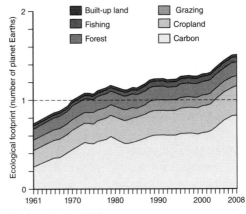

Source: *Living Planet Report*, 2012

global ecological footprint mean that future generations will face a standard of living that will be lower as a result of this generation's profligacy.

The more cheerful finding from Figure 5.1 is that, excluding CO_2, the world is still operating within its ecological boundary. We need 'only' about 70 per cent of the world's land area to feed ourselves and provide wood and other necessities. This part of the world's footprint is growing but at a much slower rate than the forest area necessary to capture the carbon from burning fuels. If human numbers peak at 10 billion at mid-century, we might still have enough land surface to meet our needs if we can cut our CO_2 emissions dramatically. In other words, if we could avoid burning oil, gas and coal, then we might be able to stay within our ecological limits.

There's more on this issue in the chapters on food supply and energy use.

Wackernagel's institute specializes in looking at the ecological footprints of individual countries. The research shows, as expected, that richer countries typically have a much larger impact than poorer countries per head of population. The style of life of people in the US means that their footprint is several times as large as the world average. It would take the space of five planet Earths if we all lived in the way that Americans do. Figure 5.2 shows that the average individual in rich countries has a footprint much greater than the 'biocapacity' of the planet, but our ecological burden is not increasing. This finding is consistent with the previous chapter's hypothesis that the rich countries are close to 'Peak Stuff'. By contrast, middle-income countries have rapidly

▼ Figure 5.2 Global changes in the ecological footprint per person in high-, middle- and low-income countries, 1961–2008

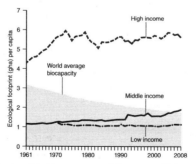

Note: A global hectare (gha) is a measure of the average biological productivity of a hectare of land across the world.

Source: *Living Planet Report*, 2012

rising ecological footprints as their economies develop. But the impact per person is still well below the level of industrial countries. Available biocapacity per person is falling because of rapidly rising population.

China's rapid industrialization has increased its ecological footprint but the level per person is still well below the figures of the richer countries. To put it crudely, the rate of increase of the global ecological footprint is being driven by the newly developing countries but its absolute level is much more driven by the much larger footprints of the EU, Japan, the US and other prosperous countries.

Wackernagel and his colleagues help countries and large companies reduce their overall ecological footprint. Others work on the footprint of individual products and services, focusing on the amount of CO_2 produced to manufacture the item. This helps manufacturers concentrate on those industrial processes that cause the greatest greenhouse gas emissions and then reduce them. Many large companies have made major reductions in their carbon footprints, partly because this saves them money, partly because of a sense of social responsibility, but also because it helps protect them from unforeseen shifts in the price and availability of energy. Measuring a footprint helps institutions manage it downwards.

6

Food

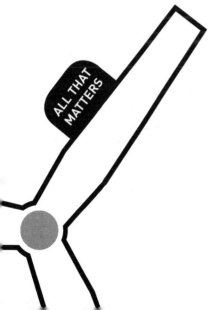

ALL THAT MATTERS

The next four chapters look at individual parts of the modern economy, examining them in the context of the central sustainability issues identified so far.

▶ The food problem

The sustainability of our food supply involves two separate questions. First, does the world have enough decent land and adequate supplies of fresh water to provide a healthy quantity of food for 10 billion citizens in 2050 and after? And, equally importantly, do modern agricultural methods impose unsustainable pressures on the world's environment? In one way or another, farming probably creates about a quarter of all the greenhouse gases produced by human activity. The use of artificial fertilizers is also responsible for ocean 'dead zones' and other ecological problems. The loss of forest to the growing of crops is detrimental to climate stability.

▶ Photosynthesis

To begin with, we need to provide a little background on why the food issue is critical to the sustainability debate.

As we have seen, the world has to survive on the raw materials that it was given when the planet was formed billions of years ago. Broadly speaking, we should be able to manage to give people a decent standard of living by using, and then continuously reusing, these materials. (Though no one would say it is going to be easy.)

In addition to our planetary endowment of materials and fuels, we get something free every day to use: almost unlimited energy from the sun. Solar radiation is vitally important in two ways:

1 It allows plants to grow and, in growing, extract carbon dioxide from the air.

2 It gives us an important source of potential power to replace fossil fuels.

This chapter focuses on the first role. Solar energy drives the process known as photosynthesis: the use by plants and trees of the energy in the sun's light to convert carbon dioxide from the air into energy-rich glucose. Glucose, a simple carbohydrate, is the building block from which all other living material is ultimately made. (There are a few exceptions to this, but we can ignore them because of their relative unimportance.)

Perhaps 2 per cent of the energy hitting the leaves of a plant in warm, fertile and well-watered soils is captured by photosynthesis. By contrast, deserts, ice caps and seas will often have few organisms able to use the sun's rays, and across the globe as a whole only approximately one-tenth of 1 per cent of the energy of the sun is successfully employed in producing glucose. Nevertheless, this is the entire basis of life on this planet. Every living creature consumes either plants or animals that have previously eaten plants. As the biblical prophet Isaiah said, 'All flesh is grass.'

The clearest illustration of the importance of solar radiation comes from the volcanic eruption of 1815 at Tambora in modern Indonesia. This massive explosion caused many millions of tonnes of fine particles to enter the atmosphere, temporarily reducing the intensity of the sun's rays reaching the Earth. This reduced temperature and also cut the amount of photosynthesis. Crops failed around the world and mass starvation resulted. Without the sun's energy, humankind would not be able to feed itself.

Photosynthesis was also the ultimate source of fossil fuels now deep beneath the ground. Many millions of years ago, plants or small organisms such as algae captured some portion of the sun's energy, grew and then died. As they decomposed into carbon and carbon-based molecules such as methane, they became coal, natural gas or oil.

We can – and do – use fossil fuels to replace some of the energy normally gained directly from the sun. Greenhouses that use man-made electricity to provide light in winter to encourage growth in crops are, in effect, doing just this. But the scale of global photosynthesis is far too great to be replaced by fossil fuels. Very approximately, 120 gigatonnes of carbon are extracted from the atmosphere each year and turned into energy – providing simple glucose molecules which are then turned into more complex carbohydrates such as cellulose. For comparison, the total amount of carbon in the fossil fuels the world uses each year is only about 9 gigatonnes.

The energy value of glucose is the amount of work it can do when it is converted back to carbon dioxide and water. The energy in food is usually referred to as 'calories', and glucose provides roughly 4 calories in each gram. Each year, a properly nourished person needs to eat the equivalent of about 220 kilograms of carbohydrates formed from glucose molecules. Multiplied by the 7 billion people on Earth, this is less than 1 per cent of the total energy captured by photosynthesis. Although most of this carbohydrate is in the form of indigestible cellulose and lignin, there is certainly no shortage of potential food supply.

▶ Farming

Put this way, it seems as though sustainability is not a problem. If the sun's energy is providing this much new food energy each year, why would we ever worry about not having enough food? Doesn't this mean that a population of 10 billion could easily feed itself?

This is not as simple a question as you might think. Today about 12 per cent of the land area of the planet is used for cultivation, growing crops as varied as rice, palm oil, oats and sugar cane. A somewhat larger percentage is available for pasturing animals for meat and for milk, but this land provides less than a quarter or so of human nutritional needs. And because almost all pasture-land is unsuitable for cultivation, being too wet, dry, cold or infertile for crops, we will have to continue to rely on cultivated cropland.

Advances in plant breeding, greater irrigation and better understanding of how to farm productively have helped the world increase the amount of food grown each year. Per head, the net availability of nutrition has risen from about 2,200 calories a day in the early 1960s to about

2,800 in 2008, even as the global population has risen sharply. (The total value of food produced per head is about 5,000 calories, but much of this is fed to meat animals.) About three-quarters of this improvement has arisen because farm yields have risen and one-quarter from extending the area of land under cultivation, principally from cutting down old forests.

In developing countries, calorie availability rose from 1,850 to more than 2,600 in the same period. The percentage of people chronically undernourished in developing countries fell from 34 per cent in the mid-1970s to 15 per cent three decades later.

The last few years have seen a small reversal of these positive trends, partly because the demand for meat in newly prosperous countries is increasing rapidly. Nevertheless, the evidence is that the world's farmers are continuing to get more tonnes of food from the world's cultivated lands every year. Global agricultural production is generally rising between 1 and 2 per cent a year, just about enough to match the increase in the number of people and to slowly improve diets. But will this continue? It depends on three things:

1 *Continuing improvement in agricultural methods*, particularly in areas of low productivity per hectare, such as much of the continent of Africa.

2 *Care of the soil.* Perhaps a third of the world's arable lands are degraded by overuse, excess levels of salt or other problems caused by poor farming practices. This degradation can usually be slowly reversed by better care of the soil.

3 *Irrigation availability.* Yields on irrigated land are typically twice the levels of rain-fed soils. But water supplies are limited and agriculture is already taking three-quarters or more of all water extracted from rivers.

Even with these worries, it seems that the world has enough agricultural land to provide food for the population of the planet in 2050 and beyond. The simple analysis below shows this by looking at how many people can live off one acre of land used to grow grains such as rice or wheat. A diet consisting entirely of grains is not properly nutritious or satisfying but this calculation gives a good guide to the biological capacity of today's agricultural land. It is a reasonable simplification because about 50 per cent of the world's consumption of calories comes from cereals.

▶ *Today's cultivated land* – about 16 million square kilometres or 1,600 million hectares.

▶ *Amount of currently cultivated land per person* – about 0.22 hectares (today, with 7 billion people) or about 0.16 hectares (in 2050, with 10 billion people).

▶ *Today's global average of grain production per hectare* – about 3.5 tonnes.

▶ *Calories produced by average grain production per hectare* – 11.5 million per year (assuming 3,300 calories per kilogram).

▶ *Area needed to produce enough calories to provide 2,500 calories per person per day* – about 0.08 hectares.

So we need 0.08 hectares (an area of just 20 metres by 40 metres) to provide sufficient calories for a person.

In theory, if we all restricted ourselves to eating grain from existing arable land, we would have twice as much as we needed in 2050, even with no further increase in yields.

Of course, it is more complicated than this. We do not just eat one thing and agricultural yields in terms of calories per hectare can be much lower for other crops. More importantly, the world's farmers divert large amounts of food production towards feeding animals and for conversion into biofuels as a replacement for oil. One-third of the total production goes to animal feed and about a fifth goes for other uses including conversion into ethanol, which can replace petrol (see Figure 6.1).

Animals also use a substantial fraction of the oils produced on farms from the growing of crops such as soy and oilseed rape. For every calorie of meat, the animal has probably ingested about 8 calories of food. Even though meat represents only about one-tenth of the worldwide intake of calories, a very large percentage – perhaps as much as 15 per cent – of the world's total biological capacity is being used to fatten animals, probably about three times as much as is used for all other foods.

The consumption of meat in rich countries is not rising from the current average of about 80 kilograms a year. However, as living standards increase in the developing world, people are eating more meat and we should probably assume that they will copy the rich world's eating patterns. A larger fraction of world agricultural output will be devoted to meat production each year.

▼ Figure 6.1 Total world grain production

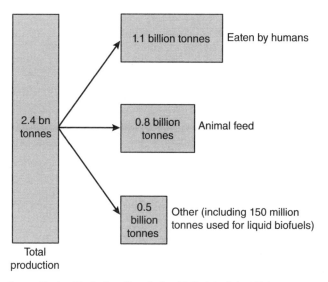

Source: Food and Agriculture Organization (FAO) of the United Nations

Will the world be able to provide a much larger population with 80 kilograms of meat each in 2050, implying a near threefold increase in animal production? No, is the very likely answer. If current patterns persist, the world would be using *all* its 2012 grain production to feed animals, and a very large fraction of all the other foods grown on Earth's surface. One recent scientific paper put it another way, suggesting that a 2050 world that has agricultural productivity as high as the US today across the whole world, but which also copies the US's meat-eating habits, would need global croplands to double in size.

ALL THAT MATTERS: SUSTAINABILITY

Most studies show that the world has relatively little land that can be brought into food production without cutting down forests. One FAO estimate suggests a maximum figure of an additional 5 per cent of today's cultivated land area, although others mention larger figures. Similarly, more land can be productively irrigated to improve yields, but the amount of freely available water is severely restricted. In several countries, such as parts of the US and China, agriculture is using more water than is available from rainfall and is running down natural stores. Irrigation will help add to food production, but the future impact may be quite limited.

At this point you may well be saying: 'But why don't we just convert some of the third of the world's land area that is forest?' The answer moves us on to considering the vital role of other forms of vegetation in capturing and storing carbon dioxide. It may surprise you to know that even the best farmland is poor at photosynthesis compared to forest land – a tropical wood will take about four times as much CO_2 out of the air as a European wheat field of similar size. So if we expand the extent of farmland by cutting down trees, we are reducing Earth's ability to recycle the CO_2 we have generated by burning fossil fuels.

And, in addition to this, the destruction of rain forest results in a large burst of carbon dioxide entering the atmosphere as the trees are burned in order to clear the land. Although the rate of global forest loss is now falling as the world wakes up to the dangers, deforestation is still responsible for about one-fifth of all current emissions. Protecting the global stocks of woodland is one of the two or three most important challenges the world faces in its struggle to reduce the rate of global warming.

▶ A burger a day?

At the crudest level, the sustainability of our food supply therefore reduces to a question of whether meat consumption can be reduced. Or, perhaps, whether meat can be produced in ways that do not use a greater land area than at present. We are the earliest stages of trying to create meat in laboratories and by 2050 it is not unreasonable to suggest that we will have steaks and chicken breast that have never been part of a living animal. This may be a repulsive idea but it would reduce the pressure on land and the inefficient diversion of useful plant food to feed livestock. A second alternative, equally unattractive in my view, is the creation of giant farms built on several storeys in which animals are kept in even worse conditions than at present but in which living conditions are precisely controlled and waste products carefully recycled.

Even with these possible changes, it looks to me as though the world needs to reduce average meat consumption down to no more than half of today's rich world levels if we are to produce enough grain and other foodstuffs and avoid having to increase the land area given over to crops.

Optimists will claim that this conclusion is too severe. They point to the long-run increases in global yields, still continuing for almost all crops and most regions of the world, including the US. Perhaps the world can continue to boost food output by 1.5 per cent a year until 2050 as a result of improvements in seeds, better targeting of fertilizers, bringing unused land into cultivation and greater irrigation. There is certainly no reason to believe that the weight of food we can grow on an acre of cultivated land will not typically go up each year. This might mean that we can afford not to worry about the impact of rising meat consumption. I am highly doubtful.

▶ Other farming issues

Cows and other ruminant animals add to global warming because otherwise helpful bacteria in their guts produce hundreds of litres of methane a day. Similarly, piles of animal manure exude large amounts of this gas, which is approximately 20 times as effective at trapping heat as carbon dioxide (although it does not stay in the atmosphere anything like as long). Some studies suggest that 40 per cent of the world's total output of methane arises from the burping of ruminants.

We know that changes to the diet of cows can help reduce the amount of gas produced in their stomachs. Farmers would benefit, too, because methane (the principal ingredient of the natural gas that heats our homes) contains a large amount of energy. Burping methane means that the cow is expelling a source of useful nutrition that could have gone to making milk or muscle. There are frequent reports of breakthroughs that show how adding such things as vegetable oil, grape skins or curry spices may reduce methane output. We are yet to see much overall impact from these dietary changes.

The number of cows is edging gently upwards. From about a billion animals in 1965, the number rose by around 35 per cent to the end of the last decade. Although many newly industrializing countries rely on other animals, particularly pigs, for increasing supplies of meat, cows remain a hugely important source of global warming gases.

The fertilizers used on farmland also have two significant effects:

1 They add to global greenhouse gas emissions.

2 They create river and ocean 'dead zones' where aquatic life has ceased.

Making artificial fertilizers uses large amounts of energy, resulting in perhaps 1 per cent of world emissions of greenhouse gases. When applied to soils, a small fraction of the nitrogen becomes nitrous oxide, a very fierce climate-warming gas. In total, world fertilizer use is probably responsible for 2 to 3 per cent of all emissions. This may not seem hugely significant today, but this number is likely to grow as our need for better yields on marginal soils in developing countries requires us to use more artificial fertilizers.

'Dead zones' result from the effect of fertilizers running off into rivers and hence to oceans, as described in Chapter 4 on planetary boundaries. Devoid of life, these zones now affect large areas of the sea, including parts of the Baltic and the Mississippi Delta. Among other effects, lifeless seas reduce the amount of fish available as food.

One response to any discussion of the pervasive impact of modern agriculture on the world's environment is to recommend that a much larger fraction of the world's food should come from organic or non-intensive agricultural systems. This would also possibly reduce climate change emissions from the soil. Organic farmers avoid the use of artificial fertilizers, instead using nitrogen-fixing crops such as clover and farmyard manures to maintain fertility.

Are organic methods more 'sustainable' than conventional agriculture? The answer is probably 'no' because the total amount of food produced per hectare under organic rules tends to be lower – often sharply lower – than standard agriculture. All other things being equal, this means that we would need more of the global land area devoted to growing food than if we used conventional techniques. If we believe in the crucial importance of not expanding the total area devoted to farming, organic agriculture – however meritorious in other ways – looks a poor choice.

7

Steel

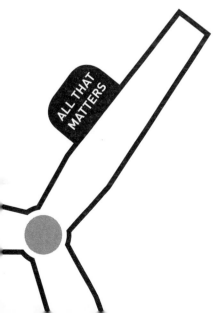

ALL THAT MATTERS

Strong, resilient and inexpensive to make, steel is the world's most important metal. Measured in tonnes of yearly production, it beats aluminium by an order of magnitude. In 2011 steelmakers made almost a billion and a half tonnes of the metal, perhaps 200 kilograms for every person on the planet. It is principally used for creating new buildings, for transport infrastructure such as bridges, for machines and for cars. Most steel is made in blast furnaces fired by coking coal, creating prodigious quantities of CO_2 emissions. Steelmaking contributes about 6 per cent of the world's greenhouse gases, the largest amount of any industrial process, though cement is not far behind.

Is steel sustainable? There are two questions here:

1 Is there enough iron ore in the Earth's crust for us to continue to make this volume of steel?

2 Can the atmosphere absorb the CO_2 produced in making steel from iron ore?

▶ Is there enough iron ore?

We're lucky that steel's key ingredient – iron – is available in such quantities. By weight, this element is about 5 per cent of the Earth's crust. Although much of this is bound up in minerals from which the metal is very difficult to extract, many rich deposits of iron oxides are dotted around the world. These oxides can be relatively easily converted to pure iron from which steel can be made.

The world will never need to use all the iron in the ground. Like most metals, steel is almost indefinitely recyclable and can be at the very centre of the circular economy. Once a building is demolished or a car scrapped, the steel is melted in an electric arc furnace

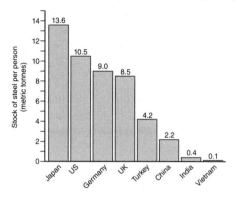

▼ Figure 7.1 How much steel does a prosperous economy need per head?

Source: Mueller, Wang and Duval, *Patterns of Iron Use in Societal Evolution* (Environmental Science and Technology, 2011)

and reformed into new ingots that have precisely the same structural strength as the virgin metal made from iron ore. Recycled cars can create new washing machines or machine tools.

There will always be small losses in the circular process, but once the world has a big enough stock of the metal, almost all of the world's requirements will be met from recycling steel that is already in existence. The evidence from advanced economies, such as the US, is that we need a stock of about 10 tonnes per person (Figure 7.1). Once a society gets to that level, its needs for the metal appear to be sated. (Japan needs more because it constructs stronger buildings to resist earthquakes.)

Although at some point in the future the world will have smelted enough ore to give each person on the planet a sufficient amount of steel, we are nowhere near at the moment. In particular, developing economies need a much greater stock of steel to give their populations a reasonable standard of living. Unsurprisingly, China is making about half the world's steel today, but within 30 years it is likely to have a large enough stock and its place as the world's largest producer will be lost to India.

By about 2100 almost all steel will be produced from existing recycled metal. The world will have come close to reaching a steady state in which the steel already made will be enough to provide for human needs. And that is before taking into account the possibility of technological change that will enable humankind to reduce its use of materials. One obvious example is cars. We use one tonne or more of steel in each automobile, but our need for improved fuel economy introduces a real pressure to decrease this weight. The average car in 2050 may not be built principally of steel, but, if it is, we can be sure it will contain less of the metal than it does at the moment.

If economic development continues and every country in the world achieves the same standard of living as in the rich West today, how much more steel will we need before taking into account the possibility of greater efficiency in use? Table 7.1 suggests a figure of about 80 billion extra tonnes, made at an average rate of about a billion tonnes a year until the turn of the next century.

Eighty billion tonnes of extra steel is going to require a large amount of iron ore to be dug out of the ground: four times as much as the world has mined since the invention of the Bessemer process for making steel in the mid-nineteenth century. Is this much ore readily available? The most widely quoted source is the US Geological Survey, which states that identified world reserves are, by coincidence, enough to provide just 80 billion tonnes of iron. This amount is estimated to be the weight of iron in identified deposits that mining

Today's worldwide stock per person (metric tonnes)	World population in 2012	Total stock of steel
2.7	7 billion or so	About **20** billion tonnes

Worldwide stock per person in 2100 (metric tonnes)	World population in 2100	Need for steel stock
10	About 10 billion	About **100** billion tonnes

	Extra steel needed	About 80 billion tonnes

companies would find worthwhile to extract using current technologies and at today's prices for ore.

If everybody on Earth gets their 10 tonnes of steel, we will use up all these high-quality bodies of ore. Many other iron deposits with less attractive characteristics exist, at least tripling the world's availability. But mining companies would probably need higher prices for their extraction to be worthwhile.

Here is a forecast for the next 100 years or so: frequent but unpredictable periods of short-term scarcity of iron ore, followed by price rises and then huge bursts of billions of dollars of investment into new but increasingly lower-grade deposits. The sharp spikes in the cost of ore will induce innovation in other materials, encouraging construction companies, machinery manufacturers and auto firms to make substantial substitutions for increasingly expensive steel. We will never run out of iron ore, but we will have to pay much more for it than we did as little as ten years ago.

What about the physical impact of mining? To produce a billion tonnes of iron every year requires two to three times as much high-grade ore. Getting this much out of the ground probably means the mining of at least ten times as much surrounding earth and rock. Surprisingly, this doesn't use a particularly large area – perhaps 10 cubic kilometres. A single mine 300 metres deep and stretching 6 kilometres in one direction and 6 in another would provide all the iron ore the world needs for a year.

Perhaps even more unexpectedly, the world already has most of the furnaces needed to take this ore and turn it into metal. The amount of new steel – excluding the metal recycled in electric arc furnaces – produced each year is already about 1 billion tonnes and this figure is unlikely to rise by more than a quarter before production peaks at some date before 2050. (This depends critically on the rate of development in the wave of countries that will follow China into industrialization.) Even before technological progress that might reduce the need for steel, the world can continue on its current path and build up sufficient stocks of the metal to give everybody enough to sustain prosperity.

In addition, we will see some reduction in the need for steel, including cuts in the weight of metal needed in cars and machinery. Although large improvements in the amount of structural metal going into buildings are unlikely, it may be that the average length of life of a steel-framed building, or one built with steel-reinforced concrete, will increase as we learn how to refurbish and extend the usefulness of older office blocks and large apartment buildings.

▶ Steel production and climate change

Making the virgin metal requires steelmakers to use large quantities of coking coal, which is almost pure carbon, to both melt the ore and to chemically react with the iron oxide. In the furnace, the carbon combines with the oxygen from the iron oxides, creating carbon dioxide and iron. Although there are alternatives to conventional furnaces, almost all the world's new steel today is made using prodigious quantities of coal. It is not easy to see how else a billion tonnes of steel is going to be made without pumping billions of tonnes a year of CO_2 into the atmosphere.

Recycled steel does not need coal, relying instead on huge electric currents flowing through the metal to melt the scrap. If we can generate all our electricity from low-carbon sources, such as nuclear power or wind, we need be less concerned about the impact of recycling steel on our CO_2 emissions.

Can the world continue to make steel and keep temperature increases to a moderate level? We can look at this question in one of two ways – the impact of steelmaking on the stock of CO_2 in the atmosphere or the percentage of allowable yearly emissions in 2050. Both come up with about the same result:

1 Making a billion tonnes of new steel a year for the next eight or nine decades will result in emissions of between 2 and 3 billion tonnes of CO_2 a year. Over the course of the next century, this will add up to about 160 billion tonnes of carbon dioxide, or about 10 per cent of what Oxford scientist Myles Allen and his colleagues think is

the 'safe' limit for how much we can add to stock of carbon dioxide in the atmosphere and still keep temperature increases below 2 degrees. (A molecule of CO_2 weighs 3.67 times the weight of an atom of carbon; 160 billion tonnes of carbon dioxide contains about 45 billion tonnes of carbon.) See Chapter 9 for more details.

2 The UK's Committee on Climate Change believes global emissions need to start falling from 2016 at 3 per cent a year. By 2050 output of man-made CO_2 and other greenhouse gases should be down to about 20–24 billion tonnes. On current trends steel will make up slightly more than 10 per cent of this.

In light of these numbers, is making new steel compatible with achieving climate change targets? In theory, yes, because 10 per cent is not a large percentage of the total budget the world has for greenhouse gases. But of course other activities will also demand their share of greenhouse gas allocations and the world may decide to give these a higher priority. Is steelmaking more important than cement production, also a major source of CO_2? Or will we prefer to allow growth in carbon emissions from the burning of aviation fuel rather than providing industrializing nations with the same stock of steel as prosperous countries already have?

The world urgently needs to agree how much of the remaining carbon budget should be allocated to such activities as steelmaking, cement production, farming and transport fuels. I write this knowing full well that such a decision will probably never happen. However, if a thoughtful world decides that projected levels of new steel production are not compatible with climate change targets we have three principal options:

1 *Capture the CO_2 from blast furnaces and store it (CCS), probably underground.* The technology for this is not yet fully developed but the task looks no more difficult than capturing CO_2 from power stations. Several of the possible technologies for capturing carbon dioxide can be used on existing steel furnaces. Unfortunately, progress on CCS is slow, partly because it will impose a substantial extra cost on steelmakers.

2 *Replace steel with other materials.* Carbon fibre is one alternative, although it is currently not recyclable. It uses far more energy than steel of equivalent strength but the source of this energy can be low-carbon electricity. The manufacturer doesn't have to use coal to make the fibres. This may be the only way that climate change objectives can be reconciled with the need for ultra-strong materials.

3 *Radically reduce the need for steel in other ways.* Rich countries already have a large enough stock of steel not to need much new tonnage, but the developing world will need huge quantities of metal. The prosperous world cannot deny poorer countries the opportunity to use steel to build high-rise urban buildings to house the billions of new inhabitants of large cities. Or stop the citizens of places such as China and India from buying cars as they become richer. Unfortunately, for most applications steel is by far the best material to use.

The sustainability challenge is to ensure that steelmaking takes place alongside carbon capture and that, insofar as is possible, alternative materials are used to replace metal. The difficulty of achieving either of these options cannot be underestimated.

8

Clothing

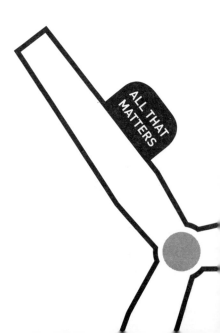
ALL THAT MATTERS

Although very different from steel, clothing manufacture also exemplifies some of the difficult ecological issues facing the planet:

▶ Crops grown as the raw materials for clothing can be enormously environmentally damaging.

▶ The manufacture of clothing is responsible for a growing fraction of total greenhouse gas emissions.

▶ Advances in manufacturing, logistics and retailing have created 'fast fashion', a trend that has pushed us towards seeing clothing as short-lived and disposable rather than items to be cared for and carefully maintained. Unbridled growth in consumption resulting from technological advances imposes unnecessary stresses on sustainability.

▶ Recycling of clothing fibres has yet to become fully established, although a fully circular economy would undoubtedly have fibre reuse as a vital constituent part.

▶ The environmental impact of cotton

The world uses about 65 million tonnes of textile fibres – artificial and natural – each year, or about 10 kilograms a person. About 35 per cent of this is cotton, with the bulk of the remainder being synthetic fibres such as polyester. Cotton is by far the most environmentally damaging fibre used in clothing. These ecological effects are principally experienced in developing countries, where most of the world's cotton is grown. However, per capita use of cotton in the form of finished textiles is four times as high in the rich world as it is elsewhere, providing the clearest example of how the environmental stresses arising from prosperous lifestyles are often transferred to poorer countries.

The slow death of the Aral Sea in Central Asia demonstrates the destructive impact of cotton farming on a regional ecology (Figure 8.1). An inland sea topped up by two major rivers, this enormous freshwater lake has shrunk by about 90 per cent as a result of the diversion of the river water to cotton irrigation. The remaining lake is too saline to support much life, destroying the local fishing industry. Much of the area that was formerly covered in water has now become a lifeless desert, with toxic dusts blown off the surface causing major problems to human health. The production of cotton in this area is one of the most destructive and unsustainable agricultural activities in the world.

In some places, cotton is grown without much irrigation, but the crop needs large amounts of water to thrive and in many parts of the world, including much of the cropland areas of India and China, a large fraction of all local fresh water is devoted to improving yields (see Figure 8.2).

In a world that is likely to be increasingly short of drinkable water, cotton is beginning to look like an unaffordable luxury. One study showed that the crop took about 2.6 per cent of the world's total availability of fresh water and in areas like the Indus river valley in Pakistan the percentage is much higher, reaching perhaps 95 per cent at some times of the year. Pakistani ground-water aquifers are also being rapidly depleted. The diversion of water that could be used for arable crops towards the

▼ Figure 8.1 Satellite photographs of the Aral Sea in 1990 (*above*) and now (*below*)

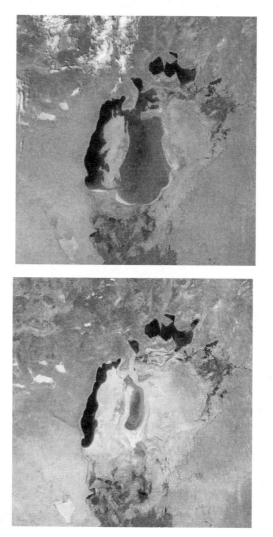

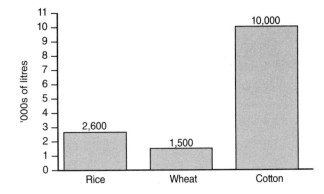

cotton fields is certain to reduce the future availability of food to meet the needs of growing populations. As the sad history of the Aral Sea shows, it is often in the areas of greatest water stress that cotton plays a central role in the local economy.

Many of those looking at the competition between food and other demands on water supply have a simple suggestion. Fresh water should have a price put on it, they say. Their arguments are strong: much irrigation is wasted and cotton farmers often have no incentive to minimize their use of valuable river water. And the eventual purchasers of the cotton, usually in the developed world, do not pay the full cost of the clothing that they buy. Do not underestimate the impact of this: the average person in the rich world buys 10 kilograms of cotton clothes a year and, at the price for fresh water in a city such as London, the cost

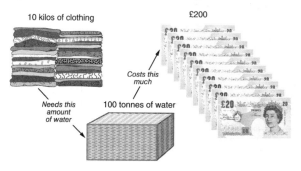

10 kilos of clothing

£200

Costs this much

Needs this amount of water

100 tonnes of water

of the water used to grow the crop would be almost £200 a year (Figure 8.3). Expressed in another way, the water used per person to grow enough cotton for a few items of clothing exceeds the average water use for all activities in the typical UK home.

Any economist will tell you that putting a price on something is often the most sensible way to ensure that scarce resources are allocated to the uses that most need them. Experiments around the world have shown that cotton can prosper on much lower amounts of irrigation water than usually employed. However, the enormous difficulties of charging by the cubic metre in countries with poor infrastructure are obvious.

The second point is that most cotton is exported from the poorer half of the world to the richer. Even with high prices for water, people in prosperous areas would still be able to afford cotton, encouraging farmers to use their acres for this crop rather than for grains or other foodstuffs. This may be a rational choice for people with land, but the impact of this on the landless poor is to reduce the local supply of food. Pricing water may be the sensible thing to do, but it does not automatically help reduce the pressure on food supplies.

Water is one problem; pesticide use is another. About a quarter of the insecticide sprayed on crops around the world is applied to cotton, ten times its share of agricultural land use. These pesticides are often highly dangerous, causing hundreds of thousands – probably millions – of serious illnesses a year. In addition to the impact from acute episodes of poisoning, the correlation between long-term pesticide exposure and mental illness – including the propensity to commit suicide – appears to be strong.

Much of the damage to humans from pesticide use is caused by farm workers failing to follow the detailed safety instructions supplied with highly toxic chemicals. Unsurprisingly, illness is concentrated in countries with low levels of literacy and poor equipment, but the US Environmental Protection Agency suggests that even in the United States 10,000–20,000 doctor-diagnosed episodes of pesticide poisoning occur each year.

The heavy pesticide load applied to most cotton has caused several major clothing companies to switch to organic cotton, which is grown without artificial fertilizer or other chemicals. At first sight, this may seem to be a more sustainable alternative. As well as the obvious benefit of not exposing workers to pesticides and reducing the pollution of local streams, the avoidance of artificial fertilizers reduces the nitrogen run-off into watercourses. Biodiversity is likely to benefit, probably reducing the pest threat to other crops being grown in the local area. Organic cotton therefore seems to be a more sustainable long-run alternative to conventional growing.

As always with low-input agriculture, the issue is yield per unit of land that is cropped. Organic methods generally produce somewhat smaller amounts of cotton, although dispute rages about the degree of difference. If the world wants to continue to produce more than 20 million tonnes of cotton lint each year, and is unwilling to increase the amount of land to achieve this, then total production will tend to fall.

We need to balance this with the other benefits of organic cotton. The reduction in pesticide use is an obvious gain, as is the cut in nitrogen fertilizer use that pollutes watercourses and the sea. Organic techniques are also less likely to result in the near-permanent degradation of land quality, which is said to be an increasingly severe problem in cotton-growing areas. As with other forms of agriculture, intensive and chemically reliant farming systems, often combined with irrigation, have provided the world with increased tonnages of cotton over the past generation. But this gain may have been at the expense of the planet's longer-term farming productivity.

The right course may not be to return to the completely organic techniques employed before the advent of artificial pesticides and fertilizers. A carefully planned programme of intercropping other plants between rows of cotton, crop rotation, reduction in pesticide use, better targeting of irrigation and the use of organic manures can hugely reduce the environmental impact of cotton growing. We need hard, careful science to help us, possibly with the aid of further genetic modification of crops to reduce water need and to increase pesticide resistance and salt tolerance.

▶ Greenhouse gases and clothing

Measuring the global warming impact of clothing – usually called its 'carbon footprint' – is not the precise science that some people claim it to be. The figure you calculate depends on a number of contentious assumptions. Nevertheless, we can guess at the CO_2 implications of an item of clothing by averaging the figures that different manufacturers have estimated for their products. Cotton and wool clothing tends to have the highest impact on greenhouse gas emissions. In the case of wool, this is because of the methane emissions from sheep. Cotton clothes have large carbon footprints because of emissions produced while growing the crop and also from processing the lint to make the fibre and then the fabric. Although cotton lint is often grown in one country, turned into fibre in another and manufactured into clothing in a third country, before being shipped to a fourth, the carbon emissions from transport are unimportant compared to the impact of growing and processing the cotton.

Adding together, the emissions resulting from cotton growing, spinning the fibre, transporting it to where the fabric is made, and then manufacturing the T-shirt and getting it to a shop might produce a carbon footprint of 10 kilograms of CO_2 for a 200-gram piece of clothing. The average inhabitant of a developed country buys about 10 kilograms of cotton clothing per year, implying a footprint of about half a tonne. This is less than 5 per cent of the average Westerner's total, meaning that it is important, but less so than food or even steel.

Perhaps surprisingly, artificial fibres such as polyester probably have far lower carbon footprints. One source estimates a figure of no more than 2–3 kilograms of CO_2 per kilogram of plastic, including polyester. Europeans use less than 10 kilograms of clothes made from oil-based synthetic fibres a year and Americans use twice as much. These figures suggest a typical carbon footprint of no more than about 50 kilograms a year, less than a tenth of the impact of cotton. This may be an unfashionable conclusion, but the use of synthetics for clothing is far more ecologically sustainable than producing increasing volumes of cotton in unsuitable soils in areas where fresh water is needed to irrigate food crops.

However, not all natural fibres are as bad as cotton. Hemp, a relative of the cannabis plant, has an insignificant carbon footprint. This weedy shrub is banned in many countries, including the US, because of an entirely mistaken belief that it contains enough hallucinogenic materials to mimic the effects of the drug. Because it requires no fertilizers, is not subject to major attack by pests and has limited effect on soils, it would be hugely beneficial to world ecology to substitute it for cotton. The chance of this happening is negligible because of the irrational connection with cannabis. Bamboo, by contrast, seems a good source of fibre, and is slowly gaining in popularity as an alternative to cotton, but requires unparalleled amounts of chemicals to process into clothing.

▶ The impact of 'fast fashion'

The price of clothing has fallen in inflation-adjusted terms as retailers such as Walmart and Inditex, the owner of Zara, have re-engineered the process of designing and making clothing and then getting it into shops in Europe and America. Clothing is perhaps the most obvious example of how economic progress can make us consume more – the UK, for example, seems to be particularly subject to the addictive effect of falling prices. In the UK, the total weight of clothing sold increased by about 40 per cent between 1999 and 2007, far faster than the rate of GDP increase. Not only was clothing not 'decoupled' from economic growth, UK consumption increased at a multiple of our income gains.

The increase in UK clothing sales up to 2007 – which have since fallen sharply as a result of recession – was partly driven by declining prices but was accompanied by a change in how people felt about clothes. Clothing was no longer carefully maintained and repaired. It became almost disposable, worn a few times and then thrown away. Clothing made from natural fibres will rot in landfill, generating methane and therefore adding substantially to the global warming impact of manufacture.

To turn fashion from being a principal driver of world ecological problems towards a less destructive role requires a move away from cotton and a reduction in the total volumes of textiles produced. I wrote in the chapter on steel that giving humankind a stock of about 10 tonnes per person seems to give people a modern standard of living with good housing, cars and reasonable infrastructure in the form of offices, bridges and stadiums. Steel can be, and usually is, substantially recycled at the end of a building's long life.

Perhaps we need to think about clothing in the same way – as a stock of capital to be conserved and properly maintained. This means buying fewer, better clothes. Lucy Siegle, a journalist who writes for the UK newspaper *The Observer,* suggests that the typical Briton is buying four times as many clothes as in 1980, and I don't think that anybody contends that we lead happier and more fulfilled lives as a result of this change. Siegle tells us to 'buy slower', planning our purchases in advance and taking pleasure from the wait.

▶ Building a circular fashion industry

Much of our clothing is as recyclable as steel or aluminium. At worst, something like a cotton shirt can be chopped into tiny pieces and the fibres reused, probably with polyester threads to increase strength. A synthetic fibre can be entirely recycled and indefinitely reused. Indeed, it can be better than this: the US outdoor wear supplier Patagonia takes used soft-drink bottles and turns them into clothing, something that circular economy specialists call 'upcycling' – turning a low-value waste into a high-quality and sophisticated product.

Full-scale recycling is only one aspect of clothing reuse. Clothing can often be physically remade to create a new item. The innovative UK company TRAID operates a large number of collection points for used clothing. Products that are torn or stained are reconstructed and redesigned into new one-off pieces and sold under an award-winning fashion label. Retailer Marks and Spencer has a larger but less ambitious scheme, offering to take back one piece of clothing for every one bought. The returned garments are then resold by the charity Oxfam or their fibres are used for remanufacturing.

The recycling of metal is well established and financially advantageous. Is this true for clothing? Probably not. The raw materials are cheap (partly because in the case of cotton and wool, the environmental costs are almost entirely unmeasured) and the processes involved in reusing clothing are complex and sometimes labour-intensive. Industry leader Patagonia commits itself to repairing all its products returned to it, reselling usable items given to it and recycling all its clothing at the end of its life. This is all excellent, but Patagonia reports that it has collected just 45 tonnes of clothes since 2005, recreating 34 tonnes of new products. It is a start, but the world needs to extend this to most of the 80 million tonnes of clothing produced each year.

Synthetic fibres are generally far easier to recycle than cotton or wool. Clothes made from oil or, even better, from plastics engineered from biological materials ('bioplastics') should be seen as a large part of the answer to the sustainability questions posed by clothing.

Energy

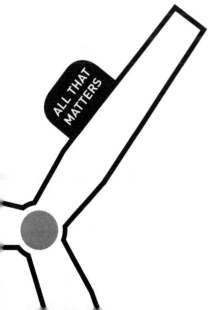

The energy problem is an incongruous one. Our planet is bathed in enough light and heat every few hours to provide all the world's power demand for a year. We may run short of copper or of good-quality agricultural land but we will never have any true shortage of energy. Nevertheless, the world's most intractable sustainability problem is climate change, deriving from our apparent inability to switch from fossil fuels to power directly or indirectly derived from the sun. Similarly, we have been able to generate electricity from nuclear fission for 50 years but electricity from this source is getting more, not less, expensive. Over the last few years, and despite the alarm felt about climate change, the world has been increasingly reliant on fossil fuels for our power. The percentage generated from low-carbon sources continues to fall.

Once in the atmosphere, additional carbon dioxide will stay there for centuries. Every tonne of coal that is burned today adds to the stock of greenhouse gases for generations to come. Yearly levels of fossil fuel use do not matter anywhere near as much as the accumulated volume of greenhouse gases that has been added in the past.

If the world chooses to burn all the fossil fuels in the Earth's crust, or even a large fraction of them, the globe is doomed to destructive global warming. This point has been made most persuasively by a team led by Professor Myles Allen of Oxford University in a short paper that contains what is now known as the 'trillionth tonne' idea. In the talks he gives to non-scientific audiences to explain this, Allen pulls out ten pieces of coal and puts them on the table in front of him (Figure 9.1). This, he says, is equivalent to the absolute

▼ Figure 9.1 The 'trillionth tonne' idea

We've burned
this much…

and can only
burn this much
more…

leaving this
unburned.

minimum estimate of the total amount of combustible carbon available to us as fossil fuels: about 5 trillion tonnes. To keep the world's average surface temperature from rising by no more than 2 degrees centigrade, only about two of these lumps of coal can ever be burned (equivalent to about 1 trillion tonnes). Unfortunately, he gloomily concludes, we have already burned one of these lumps and at today's pace, we will have burned the second within a few decades. The eight other lumps must remain in the ground for future generations to have any chance of a reasonably equable climate.

Despite Myles Allen's warning, the world's energy needs continue to trend sharply upwards and progress towards weaning ourselves off fossil fuels is very slow. While energy consumption in the developed world is largely stable, or even falling gently, newly industrializing countries are increasing their use as they become more prosperous. The International Energy Agency (IEA) suggests that the world's requirements for power are expected to rise by about one-third to 2035, a projection dissimilar to the figures provided from BP in Chapter 2. The percentage of that power coming from fossil fuels will eventually start to fall, providing 81 per cent in 2010 but falling to 75 per cent 25 years later. Even in 2035, renewable energy sources, excluding hydroelectric power, will provide only 15 per cent of the world's electricity.

Even if the commitments of governments around the world to reduce the use of fossil fuels are successfully achieved, the most likely long-term temperature increase is 3 or 4 degrees Celsius. Unfortunately, many of these policy promises are little more than pious hopes, highly unlikely to be achieved in practice. More likely, we will see even higher levels of CO_2 increase because we will continue using fossil fuels for even longer than the optimists among policy-makers suggest.

One recent IEA estimate is that the world's current trajectory will actually produce a much higher temperature rise:

> *Energy-related CO_2 emissions are at historic highs; under current policies, we estimate that energy use and CO_2 emissions would increase by a third by 2020, and almost double by 2050. This would likely send global temperatures at least 6 °C higher. Such an outcome would confront future generations with significant economic, environmental and energy security hardships.*

The IEA is understating the problem: a 6 degree rise would be catastrophic. Even a 3.5 degree rise would have dramatic effects. Not only would some parts of the globe no longer support active agriculture, but there would be hugely increased risks of drought and of flood.

Sea level will rise substantially, threatening the large percentage of the world's population that lives in coastal cities. The cost to human welfare, particularly for poorer people in the Tropics, will be incalculably large. The complacency in wealthy Northern nations about this threat is a moral disgrace.

What do we need to do? The challenge the world faces is to move from carbon-based fuels much faster that the IEA suggests is likely to happen. The only way that this can occur is:

▌ for electricity generation to shift rapidly towards renewables, such as wind and solar PV (photovoltaic), and nuclear power (together usually called 'decarbonization' of electric power);

▌ for more of our energy needs to be met by electricity, and fewer by oil and other fossil fuels. Currently, the supply of electricity in rich countries is responsible for only about 40 per cent of carbon emissions. However, decarbonization is easier for electricity supply than it is for many other energy uses, so we need to switch cars to battery power and heating towards electric heat pumps. The task is to shift energy provision so that at least 75 per cent of total energy needs are met by electricity;

▌ by dramatically improving the efficiency with which we use remaining carbon-based fuels;

▌ by not utilizing fuels from food sources ('biofuels') to substantially replace coal, oil and gas. To do so, as Chapter 6 showed, would simply add to the climate change pressure arising from deforestation as trees are replaced with plants.

▶ Reducing the CO$_2$ from electricity generation

The world electricity industry has been reliant on a small number of large electricity plants burning coal or natural gas. It must switch as fast as possible to nuclear or renewables. This will be hugely expensive, as countries around the world are finding. At the typical fossil fuel prices of 2012, old-fashioned power stations burning coal and gas were still far cheaper to operate than other sources of generating capacity. As a rough approximation, electric power from PV, wind or new nuclear will cost about twice as much as did electricity generated by modern gas turbines in 2012.

Although low-carbon electricity is currently expensive, we can expect that cost will fall, perhaps dramatically, over the next few years. Price reductions can be extraordinarily fast: a large 5-megawatt photovoltaic farm in Oxfordshire cost about £11 million in mid-2011. A year later, the quotations for similar-sized projects around the UK were about half this price.

Complete withdrawal from fossil-fuel combustion to make electricity is perfectly possible, as well as being extremely desirable. Different countries will achieve this in different ways. For Britain, the best available renewable source of electricity comes from wind power and several commentators point to the possibility of almost 30 per cent of the UK's electricity being supplied by this source in 2020. Renewable electricity has to be supplemented by energy storage and by massive upgrading of power lines to other countries so that surpluses can be exported and power deficits remedied.

Other countries might get the majority of their renewable electricity from solar energy, from biomass or from tides and waves. Solar photovoltaic power provided 40 per cent of German electricity in the middle of sunny days in May 2012, and 20 per cent across whole 24-hour periods. Finland generates about an eighth of its power from biomass and waste. Norway gets its electricity from hydro. No country yet obtains much electricity from marine energy, but the UK could almost certainly obtain a large percentage of its needs from the swift tides around its coasts. The furious energy in the shifting waters off the north-east of Scotland alone might provide 10 per cent of the UK's power.

The destruction of the Fukushima nuclear power plant will delay the building of the new generation of nuclear power stations. Nevertheless, nuclear has a critical role in weaning the world off fossil fuels for electricity generation. In countries with limited resources of renewable energy, it is difficult to see how else emissions can be reduced. China, now almost certainly the world's largest user of electricity, intends to have at least 400 gigawatts of nuclear capacity by 2050, about a quarter of its total needs. South Korea is even more ambitious, intending to generate almost 60 per cent of its electricity from nuclear power by 2030.

France's ageing fleet of reactors still produces almost 80 per cent of its power.

The evidence is increasingly strong that the speed of decarbonization is not fast enough. Global emissions are continuing to rise, increasing by 3.2 per cent in 2011 alone, and the pace at which the world is burning Myles Allen's second lump of coal is increasing. If the world is to get the electricity it needs but fails to use nuclear or renewable sources in sufficient amounts, we quickly need to find a way of avoiding the consequences of the combustion of fossil fuels. The only realistic option is the capturing of the carbon dioxide flowing out of the exhaust chimneys of coal and gas power stations. After capture, the CO_2 must be permanently stored, probably mostly in saltwater aquifers several kilometres underground or in depleted oil and gas wells.

Scientists and policy-makers have been aware of the need to develop carbon capture and storage (CCS) capacity for at least a decade. Progress has been glacially slow, with governments and electricity utilities around the world unwilling to commit the billions of dollars needed to get the first full-scale CCS plants working. And not only are CCS installations expensive to construct, they increase the operating costs of the power station: capturing CO_2 requires extra inputs of energy, mostly to separate the gas from other elements in the exhaust flue. But if we cannot find ways of building enough nuclear capacity or installing enough renewable energy plants, we have an urgent need to invest in capturing carbon. The worldwide research and development effort must be stepped up a hundred-fold.

▶ Switching more energy uses to electricity

Electricity gives us air conditioning, fridges, lift motors and tablet computers. It can also give us heating and transport. Because electricity is (relatively) easy to make without carbon fuels, the world needs to shift from using oil for transport and gas for heating to using electric power for both.

In many industrial countries, the second most important contributor to CO_2 emissions is transport. For example, in the US the use of petrol, diesel and aviation fuel produces about one-third of all man-made carbon dioxide. Battery-powered vehicles, such as the Nissan Leaf or the Chevrolet Volt, use electricity for their primary source of power (backed up by a small petrol engine in the case of the Volt). A switch from the internal combustion engine to electric motive power is a vital ingredient of any plan to reduce a country's reliance on oil.

In colder countries, gas for home and industrial heating is an important source of energy. In the UK, a cold winter can mean that almost 20 per cent of the country's emissions arise from keeping domestic homes warm. Smaller amounts are used for offices and factories. Low-carbon heating, particularly for energy-efficient buildings, is best provided by heat pumps, which use electricity rather than gas or other fuel. Heat pumps work in the opposite way to refrigerators, heating the inside of a house by transferring warmth from the outside, even when the weather is cold. As with transport, policy-makers may choose to supplement electric heat pumps with wood or other biomass, either heating single homes or groups of houses through district heating systems.

Using electric cars and heat pumps, all powered by low-carbon electricity, will enable countries to make major reductions in total CO_2 emissions. A typical industrial country can look to eventually reduce its greenhouse gas output by at least 80 per cent in this way. This will not be easy. The electrification of transport is a particularly difficult challenge because batteries are very expensive and cars are currently limited to only about 100 miles of range before they need recharging.

▶ Improving energy efficiency

In a country such as the UK the average person benefits from about 120 kilowatt hours a day of energy in one form or another. In addition, perhaps 40 kilowatt hours is contained in the creation of things we import, such as food and table computers. We can certainly make major inroads into the UK's own energy consumption. Figure 9.2 is a simplified version of Figure 18.1 in David MacKay's remarkable and hugely influential book on energy use and provides his estimate of how much energy a fairly prosperous British person typically uses. At around 200 kilowatt hours a day, such an individual has an energy consumption of about 50 per cent more than the UK average.

The efficiency of car transport can be improved by enhancements to internal combustion engines, which turn less than 25 per cent of the energy in petrol into useful motion, releasing the rest as heat. Electric cars are far more efficient, probably converting about 80 per cent of the energy in the battery into movement. So shifting from internal combustion engines to electricity has major efficiency benefits, even if fossil fuels are used for generating the electricity. The figure in Figure 9.2 falls

▼ Figure 9.2 Energy use per person per day

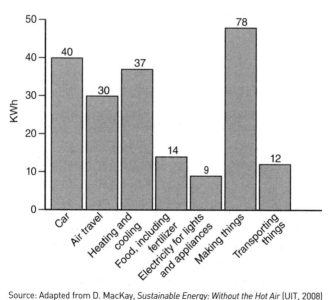

ALL THAT MATTERS: SUSTAINABILITY

Source: Adapted from D. MacKay, *Sustainable Energy: Without the Hot Air* (UIT, 2008)

from 40 kilowatt hours to about 12 kilowatt hours if the person switches to a battery-driven automobile.

Heat pumps increase the efficiency of domestic heating: they can turn one unit of electricity into up to four units of household heat. An equally important potential improvement can come through better insulation of homes, to help make them warmer in winter and cooler in summer. A heat pump might cut this well-off person's consumption from 37 kilowatt hours to not much more than 10 kilowatt hours.

Lighting and appliances take a smaller amount, and the potential efficiency savings are not as great. Moving to LED lighting and ultra-efficient appliances could reduce use from 9 kilowatt hours to about 5. As MacKay points out, the modern aircraft is already close to peak efficiency and there is little improvement to be made. But reducing the miles travelled and replacing trips with video conference calls – which I think is an efficiency improvement – is worth a reduction of at least 25 per cent in typical energy consumption.

In his book, MacKay works with the assumption that the maximum efficiency saving in other areas might be 50 per cent or so. Add all the savings together and a well-off individual might use about 60 kilowatt hours a day rather than 200 kilowatt hours. This is a really important saving, and we need to strive for cuts of this size, but it is not enough to reduce energy use to sustainable levels unless almost all sources of energy are not derived from fossil fuels.

▶ Minimizing other sources of emissions

We saw in the chapter on steel that industrial processes are responsible for several billion tonnes of CO_2 emissions per year. Apart from the refining of metals, the next most important source of carbon dioxide is cement. Fertilizer manufacture from natural gas is also an important contributor. The crucial point is this: even if we stop using fossil fuels for energy, we still face the imperative need to minimize emissions from other manufacturing and agricultural processes. This process will be significantly helped by reusing scrap materials, which generally require much lower inputs of energy than creating virgin materials. Expressed in a sentence or two, these changes sound manageable, even easy. However, each one requires sustained and painful effort over several decades.

10

An ethical challenge

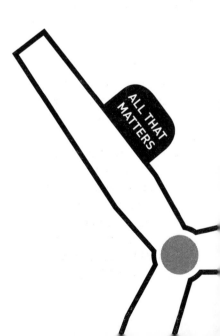

The extractive economy of the last 100 years has pulled billions out of poverty. It was almost inevitable that humankind would ignore the impact on future generations. Now we face the challenge of building a global society that recognizes the need to protect our descendants.

So far, I have not actually asked precisely *why* we need to do this. Why do we think that humankind shouldn't take out as much as it can from the Earth's crust, and do so as quickly as possible? The reason is ethical: a simple matter of fairness and equity, and nothing more. We have a moral duty to ensure that future generations – and the most vulnerable billions of our own generation – are not disadvantaged by the pursuit of easy prosperity by people in today's rich world. People put forward many different reasons for taking sustainability issues seriously but it is the moral one that is crucial.

Many of us assume that, because sustainability is an ethical issue, it must necessarily mean that we as individuals have to be the principal actors in building a longer-lasting and more equitable society. What we do in our day-to-day lives does matter enormously, particularly in the example we set other people around us. It is increasingly hard to convince sceptics that fossil fuel consumption is a serious problem if you fly across the Atlantic frequently for meetings about climate change. If you believe the future survival of human life across large parts of the world is under threat, I think it is right to accompany this belief with a commitment to a low-carbon lifestyle.

Nevertheless, we are not going to achieve much overall improvement through individual action: sustainability is actually an imperative that requires collective action across the world. That is, the world has to find a means of instigating and enforcing a set of rules on energy use and forest protection that apply everywhere. One country's actions to reduce the burning of coal needs to be matched by equal commitments in other parts of the world.

The pursuit of sustainability therefore requires planetary self-discipline on a hitherto unseen scale – those cheap seams of coal have to remain unburned, those rivers need to be left alone rather than used for irrigation, those forests that could become soy plantations need to stay woodland.

▶ What sustainability is not

As the author Mark Lynas and others have pointed out, we are well past the point at which a sustainable society could possibly be built by going back to the production and consumption habits of a long-previous era. The 7 billion people on the planet need food, housing, clothes, electricity, health care and other reasonable necessities. Quite reasonably, any so-called 'sustainable' approach that reduces the capacity of the planet to provide these services to the less well-off of today, and to the people of the future, will be rejected by almost everybody.

This book has tried to show that the world possesses the capacity and the natural resources to build a global society that can provide a decent standard of living to all. But getting to that point will require a long and difficult commitment to using science and technology to significantly reduce human impacts on the environment as we spread the benefits of prosperity to all. In my view, there is no alternative to this approach.

Many environmentalists think differently, saying that sustainability can be generated only by an immediate and permanent reduction in consumption. They say that stability and long-term maintenance of the global environment is incompatible with increasing GDP. They believe we need to return to the per capita consumption habits of perhaps a century ago. This would take us back to within the planet's limits, they say.

What was the world of 1900 like? In the case of the UK, energy use per head was only about one-third below today's levels, at least according to one detailed academic analysis. Even more strikingly, almost all of this energy came from coal with its enormous impacts on air quality, water pollution and greenhouse gas emissions. Industrial processes were highly inefficient and homes badly insulated. US energy use per head at the end of the Second World War was also just one-third lower than current levels. The impact of decades of growth since then has not been negligible, but a return to the patterns of the past would not reduce our climate change impact by a sufficient amount.

Food supply was grown in 1900 without artificial fertilizers but required far more land area per head of population. The amount of agricultural land in England fell by 12 per cent between 1900 and 2008, and the area under woodland rose. In the same period, wheat yields per hectare increased fourfold from 2 to 8 tonnes, providing more far more grain per head than a century or so ago, even though the UK population has increased by nearly 60 per cent. One of Johan Rockström's planetary boundaries is the percentage of land used for agriculture. Overdeveloped, densely populated Great Britain is the last place on Earth that Rockström might have expected to see improvement. But the application of science has brought this about.

So I suggest that any hypothesis that we need to revert to the consumption habits of the past falls at the first hurdle. If there was ever a golden age of sustainable living, it must have been before the invention of the steam engine and the huge expansion in the use of coal. There is barely a person alive who would want to go back to life before proper drainage, antibiotics and year-round availability of fruit.

Will further economic growth necessarily involve greater abuse of Earth's natural endowment? I have tried to show that the answer is not as obvious as we might think. Growth tends to make us more efficient in the use of resources and more skilled technically at providing for human needs with less impact on the environment. Invention and innovation are far more likely in a changing and flexible economy. In time, and perhaps already, material needs are fully met and growth arises from non-resource-using service activities such as education, digital entertainment and advanced telecommunications.

▶ Population

Any discussion on sustainability rapidly turns to population growth and its impact on resource consumption. At the simplest level any assessment of the effect of rising populations on sustainability has to conclude that the world would find it easy to reduce its overall ecological footprint if it contained fewer people. It would be better if population stabilized at today's 7 billion than continued rising. Fewer people means less food to be grown, less steel smelted for buildings and less oil burned for transport, all of which would help maintain environmental stability.

Hardliners tend therefore to say that the world needs to take measures to enforce a slower growth rate of population (although they very rarely say how such policies might be enforced). By contrast, writers such as Fred Pearce have commented that economic development is perhaps the best way of reducing birth rates. As countries grow, their birth rate tends to fall sharply, usually to levels little higher than the rate sufficient to maintain stable population. The impact of the rapid economic growth of the last 30 years on population increases has been profound. The percentage growth rate of world population peaked in the 1960s and has fallen sharply since. The year of the maximum net growth of population – 87 million new souls – was 1987 and according to Pearce this figure will slide towards a negative number by the middle of this century.

What is the importance of this? The optimists say that if we allow economic processes and growth to continue, population growth will cease to be a problem. It is poverty that causes birth rates to remain high and the quickest way to population stability, other than war or starvation, is a reasoned and sensible pursuit of economic growth that boosts the income of the less well-off. In a sense, this is the concomitant of the 'Peak Stuff' theory I addressed earlier: rather than assuming that economic development is the key problem, we can hope that, once society has become reasonably prosperous, its population and its material consumption will stabilize. Economic progress encourages the process of technical innovation that reduces waste and encourages energy efficiency.

This is a highly contentious and unconventional proposition. Almost everybody working in the field of sustainability regards it as dangerous nonsense. All I ask is that you keep an open mind about this issue rather than automatically siding with the view that sustainability and continued development are necessarily incompatible.

▶ The techno-optimists

There are many people who profoundly disagree with the core ideas discussed in this book for the opposite reason to the anti-growth activists. Their claim is that the best way of creating enduring world prosperity is for countries to pursue their own economic self-interest, even at the expense of greater pollution or the rapid depletion of minerals and other natural assets.

This is not necessarily a stupid or selfish argument from the techno-optimists. A poor society does not have the means to improve its environmental standards. Its people are too busy struggling to keep themselves and their families free of hunger, disease and other threats. The countries of seventeenth-century Europe may have had limited adverse impact on the natural environment but they were yet to free their populations from the threat of famine. In what sense were these countries sustainable, the techno-optimists pointedly ask, if a few weeks of bad harvest weather could cause millions to face starvation over the following winter?

As societies develop economically, the plain truth is that they do become more environmentally robust. With very few exceptions, they also take more care of their natural environment. Gross pollution, such as the discharge of wastes into water courses, becomes rarer. Wealthier countries are usually also less wasteful of resources. For example, the best practice in the making of a tonne of new steel involves the use of more than 600 kilograms of coking coal. One hundred years ago, it would have been twice this level. Perhaps even more tellingly, making steel in China today is typically about half as energy-efficient as the best plant in the West, but the difference is narrowing rapidly as Chinese manufacturers copy and improve upon the technologies available elsewhere.

Put at its fiercest, the argument of the techno-optimists and proponents of the unimpeded free market is that policies that appear to be enhancing sustainability are actually reducing economic growth, and delaying efficiency improvements. They contend that since economic growth tends to reduce environmental stresses the particular pursuit of sustainability is counter-productive if it holds back GDP increases.

'Up to a point', most environmentalists respond. Yes, richer societies do tend to become better at looking after their own backyards. Comfortably-off citizens do not want plumes of toxic chemicals erupting nightly from the local factory. But the well-off also become more ruthlessly exploitative of the wider world. For example, Europeans spent most of the last half-century plundering the fish stocks of the North Sea to the point where some species almost disappeared. Economic growth does nothing to protect the global 'commons', the resources potentially available to all of us but which no one person owns.

As importantly, increases in national income have often been accompanied by greater exploitation of natural resources. Although the efficiency of use of raw materials or land has improved, world GDP growth has still increased the pressure on scarce resources, such as fresh water and the limited amount of world arable land. It has helped drive deforestation and so increased carbon emissions.

This is the first part of the argument against the free-market optimists. The second is that, although freely functioning markets are excellent at reducing waste (because it costs money) and improving efficiency (meaning that companies can compete more successfully), they also undervalue the future wellbeing of the next generation. In a survey by the UK government published in 2008 almost 80 per cent of people interviewed expressed a preference for £1,000 today rather than £1,100 one year in the future. By extension, they would rather extract scarce resources, such as fresh water from the underground aquifer, than store it up for next year or, even more remotely, for their children's children.

Most people react with horror to the idea that today's consumption patterns might reduce the standard of living of their descendants, but the unfortunate truth is that humans do not value the future very highly. Government policies, many of which will have to be implemented across the world to be effective, and which will have to persist for many decades, will be needed to create the right incentives for a sustainable world. Free markets will not deliver it.

People, businesses and economies react to signals provided by prices. But in the case of the two central problems identified in this book, climate change and overuse of agricultural land, there are no obvious signals to respond to. As the CO_2 concentrations in the atmosphere pass through 400 ppm sometime in the next couple of years, there will be no entrepreneurs adjusting their behaviour to reflect this change. The financial advantage of burning fossil fuel and pumping carbon dioxide into the open air rather than sequestering it will always remain. Nor will agriculturalists stop cutting down virgin forest in order to increase food production. In fact, the reverse is more likely as climate change reduces yields in many parts of the world, increasing the incentive to turn woodland into arable land. The operation of the price mechanism cannot replace concerted international action on the critical questions touched on in this book.

This 100 ideas section gives ways you can explore the subject of sustainability in more depth. It's much more than just the usual reading list.

1 What does sustainability mean?

1   In the first chapter, I look at the most obvious cases of activities, such as the overexploitation of fresh water, that deplete 'natural capital' and are therefore unsustainable. A good paper on some of the problems and concerns with aquifers and surface water can be found at http://pubs.usgs.gov/circ/circ1139/. Entitled *Ground Water and Surface Water: A Single Resource*, it goes into clear detail on water supply issues.

2   The core ideas on natural capital are very well expressed by Robert Costanza and Herman Daly in *Natural Capital and Sustainable Development,* an article available at http://www.biology.duke.edu/wilson/EcoSysServices/papers/CostanzaDaly1992.pdf

3   **Overfishing** is a class of what are called 'global commons' problems. Climate change is another. A good analysis by Scott Barrett can be found at http://yaleglobal.yale.edu/content/regulating-global-commons-%E2%80%93-part-i

4   The classic statement of the **commons problem** was written by Garrett Hardin in 1968 and is available at http://www.garretthardinsociety.org/articles/art_tragedy_of_the_commons.html. (The word 'tragedy' in this phrase refers to the gradual evolution of a problem towards its inevitable unhappy consequence, rather than the way we normally understand the word.) Although the paper is more than 40 years old, it remains utterly central.

5   Later in the book, I say that **abuse of the Earth's land surface** is one of the two most intractable issues we face. The destruction of forests for agriculture, overgrazing and misuse of irrigation water are the central problems. Lecture notes on land degradation from the University of Michigan provide an introduction to the issue. See http://www.globalchange.umich.edu/globalchange2/current/lectures/land_deg/land_deg.html

6   Placing little value on the future environment is a crucial third cause of unsustainable behaviour. For non-economists, the importance of this class of problem may not be obvious. An Australian paper provides a clear exposition that should be understandable to a general reader: '**An Economic Perspective on Land Degradation Issues' by John D. Mullen** (http://www.dpi.nsw.gov.au/research/economics-research/reports/err09). Put at its simplest, if a farmer does not value the future very highly, she will overexploit the soil, trying to get as much production as possible, as quickly as possible. In contrast, someone who values farm productivity in 50 years will look after the soil with more care, even if it means immediate yields are lower. Economists think hard about how humankind discounts the future, often in strikingly different ways to ecologists.

7 For a word that seems to have an obvious meaning, 'sustainability' is remarkably difficult to define. A listing of some alternative meanings, particularly in the context of agriculture, is provided by the University of Reading at http://www.ecifm.rdg.ac.uk/definitions.htm

8 More definitions can be found at http://www.sustainable measures.com/node/36. But note, please, that many explanations drift unthinkingly into ethics. For instance 'sustainable development ... [is] the process of building equitable, productive and participatory structures to increase the economic empowerment of communities and their surrounding regions'. This is some way beyond what I think sustainability is.

9 The classic book on how some societies fall catastrophically apart because of a failure to understand the unsustainability of what they are doing is Jared Diamond's *Collapse* (Penguin, 2011).

10 Another piece of vital background reading is the 1972 Club of Rome report entitled *Limits to Growth.* This work suggested (to a generation unused to the idea of finite natural resources) that economic development might exhaust the capacity of the planet to provide goods and services. The value of this book was its use of rigorous and quantitative analysis to address environmental issues. Whether it was right or wrong is less important than the intellectual structure that it provided.

2 Are we going to run out of anything?

11 BP's yearly *Statistical Review of World Energy* is a mine of information. The 2012 edition is at http://www.bp.com/ statisticalreview/. Some people say that BP gives an unfairly rosy picture of the size of world energy reserves.

Others say that the company has been even-handed in its calculation of fossil fuel availability.

12 www.BP.com is also perhaps the best place to get an assessment of how the world will choose to provide for its energy needs in 20 years. BP thinks that fossil fuels will still be the dominant source of energy. Global carbon emissions will rise sharply. This view is not an unusual one among business and contrasts sharply with the optimism of most policy-makers. But it is also important to note that all of the increase in fossil fuel use occurs outside the currently developed world. See http://www.bp.com/liveassets/bp_internet/globalbp/STAGING/global_assets/downloads/O/2012_2030_energy_outlook_booklet.pdf

13 The US Environmental Protection Agency has a good introduction to climate change at http://www.epa.gov/climatechange/basicinfo.html. This is an even-handed analysis of the reasons why humankind's actions will substantially increase global temperatures and cause other unmanageable problems.

14 The International Energy Agency was set up after the severe oil supply problems of the 1970s. It provides governments with information and advice about energy supplies. Until recently the Agency projected continued increases in oil production but, as supplies have tightened, it now sees world oil output as only rising gently and prices remaining high. Its views on climate change can be found at http://www.iea.org/topics/climatechange/

15 The work of Yale professor Tom Graedel is central to the debate on the volume of metals needed to sustain a prosperous society. In 'Metal Stocks and Sustainability' (http://www.pnas.org/content/103/5/1209) he suggests that 'there is no immediate concern about the capacity of mineral resources to supply the requirements for geochemically scarce metals' such as copper. But he also suggests that to provide the entire world population

with as much copper as is currently used in the US would exhaust all available resources. Other metals are generally in more abundant supply.

16 *Sustainable Materials: With Both Eyes Open* by Julian Allwood and colleagues at Cambridge has a chapter on cement. It is available as a free download from http://withbotheyesopen.com/. You may need to print it out to follow the dense chain of argument. The majority of the book is about steel, which is covered in this book in Chapter 7.

17 The US Geological Service (USGS) provides a comprehensive database of information on world reserves of metal ores and other minerals.

18 Geoscience Australia offers assessment of mineral reserves both in the country and in the world at large. Because Australia is such an important producer of many metal ores and also of coal and other minerals, the documents available at http://www.ga.gov.au/minerals.html are vital sources of commentary on likely trends in mineral availability.

19 A closer look at reducing steel use than is possible in this book would include sections on the use of carbon fibres as a way of avoiding the emissions of CO_2 during manufacturing processes. Because carbon fibre is very difficult to recycle, and takes a great deal of energy to make, carbon fibre does not look like an easy replacement for steel. However, it is very light and 'lightweighting' electric cars will improve the number of miles they will travel on one charge. *The Economist* published a good article on the car industry's approach to the use of carbon fibre: http://www.economist.com/blogs/newsbook/2011/03/vw_buys_bmws_carbon-fibre_dream

20 Carbon capture of the CO_2 from blast furnaces is a potential alternative means of meeting emissions targets.

Progress towards, this is painfully slow, and few see the potential for any working storage projects before 2030 or later. The industry association looking at the possibility is called Ulcos and its website is at www.ulcos.org

3 Will economic growth make our problems worse? An unconventional view

21 Without careful quantification of humankind's use of resources, no one can know the seriousness of the sustainability problem. Fridolin Krausmann's research team has provided both the conceptual foundation and accurate numbers for many of the world's economies. *The Global Socio-metabolic Transition: Past and Present Metabolic Profiles and Their Future Trajectories* is a key source of information. Downloadable from http://www.uni-klu.ac.at/socec/downloads/2009_KrausmannFischer-KowlaskiSchandl_metabolic_transition_JIE12_58.pdf

22 Fridolin Krausmann edited a short book on different national trends in resource use, including chapters on India and Japan, entitled *The Socio-metabolic Transition.* It can be read at http://www.uni-klu.ac.at/socec/downloads/WP131FK_webversion.pdf

23 The Institute of Social Ecology at the University of Klagenfurt in Austria (helped by Statistics Austria) produced a report on material flows inside the Austrian economy that is exceptionally easy to read. Entitled *Resource Use in Austria, Report 2011* (Austrian Federal Ministry of Agriculture, Forestry, Environment and Water Management), it provides a model for measurements of national resource use.

24 In late 2011 I wrote a paper entitled 'Peak Stuff' showing that most indicators of the material needs of the UK economy suggested a peak around the turn of the millennium

and a decline since then. This can be found at www.
carboncommentary.com/wp-content/uploads/2011/10/
Peak_Stuff_17.10.11.pdf

25 *Prosperity Without Growth* by Tim Jackson (Earthscan, 2009)
very quickly became the standard text on why economic
growth is bad for people in developed societies. Jackson
is strongly of the view that economic growth remains
correlated with increases in material consumption. He
advocates building a society that focuses not on economic
growth but on improving the efficiency of resource use.

26 Tim Jackson wrote an article on why my 'Peak Stuff' theory –
even if true – was not helpful. See http://www.guardian.
co.uk/environment/2011/nov/01/peak-stuff-message-
green-technology. Michael Blastland compiled a piece about
material flows for the BBC website, also sceptical about
'Peak Stuff'. See http://www.bbc.co.uk/news/magazine-
15863049

27 In one of his insightful pieces on the 'Peak Stuff' idea, Fred
Pearce contrasted what might be happening in the UK
with the continuing rapid expansion in Chinese materials
use. See http://e360.yale.edu/feature/the_new_story_of_
stuff_can_we_consume_less/2468/

28 In the US, the leadership on material flows analysis has
been provided by the World Resources Institute (WRI).
*Material Flow Accounts: A Tool for Making Environmental
Policy* (WRI, 2005) gave the reasons why US policy-
makers should choose to pay close attention to tonnages
of materials entering and leaving their economy. It is as
vital to environmental policy as GDP measures are to
economic policy.

29 Material accounts for the US were published by WRI in
2008. See *Material Flows in the United States: A Physical
Accounting of the US Industrial Economy* at http://www.wri.
org/publication/material-flows-in-the-united-states

30 The EU's careful description of how material flow accounts work is provided on the Eurostat website. In simple language, the web pages show the importance of this form of accounting. See http://epp.eurostat.ec.europa.eu/statistics_explained/index.php/Material_flow_accounts

4 Planetary boundaries

31 The full Planetary Boundaries paper is available from The Stockholm Resilience Institute at http://www.stockholmresilience.org/download/18.8615c78125078c8d3380002197/ES-2009-3180.pdfs

32 Johan Rockström's speech at a TED Oxford conference can be seen on the TED website at http://blog.ted.com/2010/08/31/let-the-environment-guide-our-development-johan-rockstrom-on-ted-com/

33 The environmental writer Mark Lynas was invited to help the planetary boundaries scientists explain their ideas to the widest possible audience. In *The God Species* (Fourth Estate, 2011), Lynas wrote an impassioned statement of the reasons why humankind should work to understand the natural boundaries within which we all live.

34 Some of the reviews of *The God Species* give reasonable summaries of the importance of the boundaries concept. The *Guardian* article is at http://www.guardian.co.uk/books/2011/jul/20/mark-lynas-god-species-review

35 Marek Kohn's somewhat tart review of *The God Species* in the *Financial Times* provided another perspective: http://www.ft.com/cms/s/2/cb6cb66e-a976-11e0-bcc2-00144feabdc0.html#axzz1x90oj4hk

36 Andrew Balmford et al., 'What Conservationists Need to Know about Farming', *Proceedings of the Royal Society B*, 25 April 2012, suggests that the boundaries related to agriculture may operate more at a very local scale rather than at a planetary level.

37 The most acerbic criticisms of the 'planetary boundaries' idea, and its value in making future global policy, has come from the Breakthrough Institute, a Californian think tank. Its paper 'The Planetary Boundaries Hypothesis: A Review of the Evidence' concludes that the Rockström argument that natural systems that move outside boundaries will 'tip' into dangerous states is backed up by little evidence except in the case of climate change.

38 A similarly unconvinced view about planetary boundaries was voiced by *The Economist* – see http://www.economist.com/node/21556897. The magazine makes the point that only three of the boundaries really have thresholds in any substantial sense: ozone, ocean acidity and greenhouse gases.

39 A powerful assessment of the evidence for tipping points in global ecological systems was published in *Nature*. Anthony D. Barnosky and his colleagues examine the plausibility of rapid planetary-scale shifts of ecological state, concluding that 'biological resources ... may be subject to rapid and predictable transformations within a few human generations'. See http://www.nature.com/nature/journal/v486/n7401/full/nature11018.html

40 The editors of *Nature* magazine provided a very good summary of the Barnosky article. 'Most forecasts of how the biosphere will change in response to human activity are rooted in projecting trajectories. Such models tend not to anticipate critical transitions or tipping points, although recent work indicates a high probability of those taking place. And, at a local scale, ecosystems are known to shift abruptly between states when critical thresholds are passed. These authors review the evidence from across ecology and palaeontology that such a transition is being approached on the scale of the entire biosphere. They go on to suggest how biological forecasting might be improved to allow us to detect early warning signs of critical transitions on a global, as well as local, scale.'

5 Two central ideas

41 A 1982 paper by Walter R Stahel has a claim to be the first attempt to show the environmental importance of building durable, reusable and recyclable products. The paper – 'The Product-Life Factor' – can be found at http://product-life.org/en/major-publications/the-product-life-factor. There is lots of other interesting material on this website, though it is often technical.

42 *McDonough and Braungart's Cradle to Cradle* (2002) was the first book to popularize the advantages of continual reuse of materials. A quick summary of the core ideas is provided in an interview with Michael Braungart on the *Forbes* magazine website. See http://www.forbes.com/sites/terrywaghorn/2012/03/23/william-mcdonough-remaking-the-way-we-make-things/

43 Ellen MacArthur, the former ocean yachtswoman, set up a foundation to encourage businesses to use 'circular' manufacturing methods. Her organization's report at http://www.thecirculareconomy.org/exec-summary is a good summary of how companies can move from the old model of making things that quickly turn into useless, and possibly toxic, waste to manufacturing objects that become 'nutrients' for the next generation of valuable products.

44 Ronald Clift and Julian Allwood wrote a short article about the possible conflicts between the standard quasi-capitalist economic model and circular approaches to manufacturing. It can be found at http://www.ellenmacarthurfoundation.org/circular-economy/circular-economy/rethinking-the-economy

45 Carpets are resource-expensive to produce and are usually disposed of to landfill. But they can be almost wholly recycled back into new carpets. An inspirational case study on Desso Carpets from the Netherlands is very well worth reading. See http://www.ellenmacarthurfoundation.org/business/featured-articles/desso-10-years-to-close-the-loop

46 Another exemplary company is **furniture manufacturer Herman Miller**. An example of how closed-loop thinking permeates the operations of this company can be found at http://www.hermanmiller.co.uk/publicwebresources/documents/products/seating/celle/celleEnvironmental_Summary_Sheet.pdf. The document describes the environmental credentials of its Celle chairs in detail.

47 A good description of the '**Global Footprinting**' idea can be found at http://www.footprintnetwork.org/en/index.php/GFN/page/basics_introduction/. Although some writers have criticized the concept because it combines greenhouse gas sequestration with the wholly different idea of resource use, the idea has appeal as a single memorable index of sustainability.

48 **Best Foot Forward** is a leading proponent of ecological footprinting for companies and other organizations. The value of work for customers such as the London Olympics is that they get a measure of the current sustainability of their operations. Best Foot Forward then produces plans for the reduction of those impacts. Case studies can be found at http://www.bestfootforward.com/casestudy/

49 The most readable guide to **carbon footprinting** for non-specialists is Mike Berners-Lee's *How Bad Are Bananas* (Profile Books, 2010). One of Berners-Lee's strengths is a commitment to rigorous quantification accompanied by a focus on the issues that really matter.

50 *How to Live a Low-carbon Life* (2nd edition, Earthscan, 2010) is my attempt to show the carbon footprint of the major human activities, including household heating, driving and eating.

6 Food

51 A University of Michigan lecture series has notes that provide an utterly superb introduction to the idea of 'net primary production' (NPP) and how it is allocated. See http://www.globalchange.umich.edu/globalchange1/current/lectures/kling/energyflow/energyflow.html

52 To answer the question 'Is there going to be enough NPP to feed people?', see http://www.eoearth.org/article/Global_human_appropriation_of_net_primary_production_(HANPP)

53 Twelve per cent of the world is cropland. This article demonstrates why it may not be possible to increase this percentage without reducing the ability of the biosphere to capture CO_2 from the atmosphere. See Peter Vitousek et al., *Human Appropriation of the Products of Photosynthesis* at www.jstor.org/stable/131028

54 Some parts of the world's surface generate more net primary production than is consumed by the human population. Others are in deficit. Marc L Imhoff et al., 'Global Patterns in Human Consumption of Net Primary Production', *Nature* (2004) at http://www.fraw.org.uk/files/economics/imhoff_2004.pdf provides an appraisal.

55 Thomas Kastner and his colleagues show that the growth of world meat consumption will soon become a more important contributor than population growth to the global requirements for arable land. See *Global Changes in Diets and the Consequences for Land Requirements for Food* at http://www.uni-klu.ac.at/socec/downloads/Kastner_et_al._-_2012_-_Global_Changes_in_Diets_and_the_Con_2.pdf

56 *The FAO Statistical Yearbook* provides detailed information on world agriculture, including detailed analysis of land use and productivity.

57 Chapter 1 of *The State of the World's Land and Water Resources for Food and Agriculture*, published by the FAO at http://www.fao.org/nr/solaw/thematic-reports/ en/ assesses trends in the use of global land and water resources.

58 Chapter 2 of the same report looks at the competing uses for **the shrinking amount of spare land.** See http://www. fao.org/nr/solaw/thematic-reports/en/

59 This chapter has been unsupportive of organic agriculture, though personally I would prefer if agriculture adopted **organic techniques** such as the use of clover to build soil fertility. A good analysis of the climate change impact of organic agriculture can be found on the website of the Institute of Science and Society at http://www.i-sis.org.uk/mitigating ClimateChange.php

60 **William Little** wrote a fascinating article on **lab-grown meat** in the *Financial Times* of 21 April 2012. See http:// www.ft.com/cms/s/2/87bf2654-89b3-11e1-85af-00144feab49a.html#axzz1xZgbybc5

7 Steel

61 **Is sustainability essentially an engineering problem?** One that can be solved by cold-blooded analysis? Or is it fundamentally an ethical question about how best to rein in mankind's self-destructive tendencies? The Department of Engineering at Cambridge has, as you might expect, the former view. Its 2011 book *Sustainable Materials: With Both Eyes Open* (http://withbotheyesopen. com/) is an analysis of how high levels of materials use can be made compatible with a sustainable world. This chapter makes very extensive use of this thoughtful and rigorous book.

62 The literature on sustainability is riddled with the **incorrect assumption that human needs are infinite.** A core purpose of this chapter is to show that for most materials we can estimate the total stock that the world requires to meet the maximum human requirements. A useful paper on estimating world needs is 'Patterns of Iron Use in Societal Evolution' by Daniel Mueller et al., published by Environmental Science and Technology in 2011. See http://pubs.acs.org/doi/pdfplus/10.1021/es102273t

63 **Tom Graedel's paper on 'In-use Stocks of Metals: Status and Implications', written with Michael Gerst** in 2008, provides an estimate of how much metal the world has in use and how much more will need to be refined to give everybody the stock they need. See http://pubs.acs.org/doi/abs/10.1021/es800420p

64 **Graedel's article 'Getting Serious about Sustainability' (with Robert J Klee)** provides a clear methodology for assessing the required stock of a metal and the implied rate of annual production when that stock is achieved in a particular society. See http://pubs.acs.org/doi/pdf/10.1021/es0106016

65 **The US Geological Survey** provides robust estimates of the availability of metal ores in the top few kilometres of the Earth's crust. With few exceptions, the USGS sees mineral reserves as sufficient for our needs, although the costs of mining lower and lower-grade ores will inevitably rise (encouraging recycling). The numbers are at http://minerals.usgs.gov/minerals/pubs/commodity/iron_ore/

66 The second core idea in this chapter is the importance of building a **'circular' or 'closed-loop' metal economy.** Because China is the world's largest user of iron ore by a large margin, the progress of Chinese recycling is utterly critical. The following article does the arithmetic to show how Chinese steel manufacturing will deliver a sufficient stock for the country's needs if it follows a similar pattern to other industrializing countries. 'Moving towards the Circular

Economy: The Role of Stocks in the Chinese Steel Cycle', by Stefan Pauliuk et al. is at http://pubs.acs.org/doi/full/10.1021/es201904c

67 **Most carbon dioxide is emitted as a by-product of the generation of energy**. However, in the case of the making of new steel and the manufacture of cement, the CO_2 is driven off as a result of the manufacturing process. Energy use can be 'decarbonized' by replacing fossil fuels with renewable sources or nuclear power. This is impossible with conventional manufacturing of steel or cement. The key question is: Is the world able to meet its emissions targets and still build up the stock of these products that it needs?

68 **A paper by Julian Allwood and Cambridge colleagues** looks at whether the achievement of a satisfactory stock of metals and other building blocks of the modern economy is compatible with climate change objectives. It suggests that considerable reductions in the amount of metal used to achieve our purposes will be needed if we are to reduce emissions from manufacturing by 50 per cent by 2050. 'Options for Achieving a 50% Cut in Industrial Carbon Emissions by 2050' is at http://pubs.acs.org/doi/abs/10.1021/es90290

69 **Carbon fibre** does not look like an easy replacement for steel. However, it is very light and 'lightweighting' electric cars will improve the number of miles they will travel on one charge. *The Economist* published a good article on the car industry's approach to the use of carbon fibre. See http://www.economist.com/blogs/newsbook/2011/03/vw_buys_bmws_carbon-fibre_dream

70 **Carbon capture of the CO_2 from blast furnaces** is a potential alternative means of meeting emissions targets. Progress towards this is glacially slow, and few see the potential for any working storage projects before 2030 or later. The industry association looking at the possibility is called Ulcos and its website is at www.ulcos.org

8 Clothing

71 Kate Fletcher and Lynda Grose, *Fashion and Sustainability: Design for Change* (Lawrence King Publishing, 2012) tries get the fashion industry to engage with the ethical issues that clothing raises. Fletcher's blog at http://www.katefletcher.com is full of thought-provoking articles on sustainability.

72 Alison Gwilt and Timo Rissanen, *Shaping Sustainable Fashion* (Earthscan, 2011) looks at how the clothing industry can reduce its environmental impact, from pattern design through to reducing requirements for laundering. In the final chapter Kate Fletcher says that we need to become more 'materialistic', gaining contentment from investing almost sacred meanings in the things we own, rather than believing that acquisition of goods is the way to happiness.

73 Lucy Siegle rails at the textile industry for its ethical failings in *Fashion to Die For* (HarperCollins, 2011). The book covers pesticide misuse, forced labour and what is sometimes called 'the dark side of globalization'. I have tried to suggest that these hugely important ethical questions are *not* always identical to issues of sustainability but Siegle's book is an absorbing read.

74 Patagonia is the world leader in understanding the ecological impacts of textile manufacture. Until recently (mid-2012) its website had analyses of the climate change impact of representative items of its clothing. These calculations, called 'The Footprint Chronicles', may have returned by the time you read this book.

75 Available from MCL Global, Simon Ferrigno's very short book *Cotton and Sustainability* looks at many of the ecological issues involved in cotton farming, dispassionately analysing ways of reducing the impact of cotton growing on the communities reliant on this crop.

76 **The research report of Julian Allwood and his team,** *Well Dressed: The Present and Future Sustainability of Clothing in the United Kingdom* (Cambridge University Institute for Manufacturing, 2006) quantified the impact of textile manufacture on the environment. Some of the data in this report is now a little old but the working methods are exceptional. Allwood was a key player in *Sustainable Materials*, featured in the chapter on steel. See http://www.ifm.eng.cam.ac.uk/sustainability/projects/mass/uk_textiles.pdf

77 **The Better Cotton Initiative** at www.bettercotton.org seeks to decrease the environmental impact of cotton by trialling different production methods. Pilot results from China and other major producing countries should now be available on its website.

78 **H&M** has a deep commitment to building a more sustainable supply chain, including through the use of organic fibres, and its annual report on the subject contains thorough research and useful commentary on other ecological issues. See http://about.hm.com/content/hm/AboutSection/en/About/Sustainability.html

79 At the other end of the size spectrum, **TRAID's 'upcycling' of worn clothes** is an exemplary and inspiring business. See http://www.traid.org.uk/how.html

80 **Marks and Spencer's Plan A** is one of the most ambitious carbon-reduction projects in the world. It also focuses on the wider environmental impacts of the business. One of Plan A's features is 'Shwopping', a scheme to reduce the volume of clothes going to landfill. More on Plan A can be found at http://plana.marksandspencer.com/

9 Energy

81 The best introduction to climate change is **Mark Lynas's** **award-winning** *Six Degrees* (HarperCollins, 2007). This book takes the reader on a journey, degree Celsius by degree Celsius, as the planet heats up. Rooted in mainstream climate science, Lynas shows the catastrophic effects of a temperature rise greater than 2 degrees by looking at periods in the long-distant past when the climate was hotter than today. By the time the book has reached 6 degrees, humankind's entire existence seems to be in doubt.

82 **David MacKay's book** *Sustainable Energy: Without the Hot Air* is available in printed form or as a free download from http://www.withouthotair.com/. MacKay's beautiful techniques for assessing the costs and potential for conversion to low-carbon sources of energy will never be bettered.

83 More technical analysis of the costs and consequences of renewable energy can be found in the works of **Godfrey Boyle**, including his *Renewable Energy* (Oxford University Press, 2004). (A third edition may now be available.)

84 **Boyle's course at the UK's Open University** has a very useful website at www.open.ac.uk/T206/Weblinks/T206links.htm

85 In *Ten Technologies to Fix Energy and Climate* (2nd edition, Profile Books, 2009, published elsewhere as *Ten Technologies to Save the Planet*), I show the width of renewable-energy technologies that are available. I suggest the importance of also drawing down CO_2 from the air in order to counterbalance the effect of continuing fossil fuel use.

86 These drawdown technologies should, in my opinion, be focused on those **measures that disrupt the carbon cycle.** (The carbon cycle is the process that sees photosynthesis absorb CO_2 into plants and trees, which eventually die, eventually returning carbon dioxide to the atmosphere.) One of these technologies is biochar, a way of sequestering carbon. Cornell scientist Johannes Lehmann gives more information at http://www.css.cornell.edu/faculty/lehmann/research/biochar/biocharmain.html

87 I also believe in the importance of restoring dry lands that have suffered loss of soil carbon because of overgrazing or other abuse. The potential to extract atmospheric carbon is gigatonnes per year. The **photographs on Tony Lovell's website in Australia** prove also that restoring good-quality vegetation on the most degraded lands is entirely possible. See http://www.soilcarbon.com.au/case_studies/pdf/08TL_SCCPPP_En.pdf

88 **The world will continue to use fossil fuels for generations to come.** The effects will be dire unless we can find ways of capturing the CO_2 from combustion. MIT specialists wrote an article some years ago about the possible technologies to be used. It is still highly relevant and can be found at http://sequestration.mit.edu/pdf/enclyclopedia_of_energy_article.pdf

89 **The Swedish energy utility Vattenfall** is a major generator of coal-fired electricity but recognizes that continuing to burn carbon-based fuels is likely to result in the most severe consequences for humanity. Its carbon capture and storage programme is among the most advanced in the world. Details about the various possible ways of extracting CO_2 and storing it underground are provided on the Vattenfall website at http://www.vattenfall.com/en/ccs/research.htm

90 **Geoengineering** is an umbrella term that covers all attempts to reduce CO_2 levels in the atmosphere or decrease the amount of solar energy reaching the Earth. The Royal Society's 2009 report details the costs and benefits of both types of approach. See http://royalsociety.org/policy/publications/2009/geoengineering-climate/

10 The ethical challenge

91 Fred Pearce points out that population issue is rapidly disappearing. Malthus was wrong: population growth is not choked off by food scarcity, but it definitely is controlled by prosperity. *Peoplequake* (Eden Project Books, Transworld Publishers, 2010) changed my views overnight.

92 'What about China?' everybody moans. Yes, **China's appetite for resources** is growing at an unmatched rate. But it is also the economy with the greatest capacity to use clean technology to reverse the growth of emissions. The website www.chinadialogue.net provides analysis of China's environmental plans and what the world can learn from them.

93 I think we should listen to those asking us to be **more nuanced in our thinking** about sustainability. The fact, for example, that copper ores are scarce does not *necessarily* mean that we should restrict our mining today to leave more for the future. We could, for example, choose to spend more on developing alternatives. The most stimulating 'sceptics' are Bjorn Lomborg (*Cool It*) and Matt Ridley in *The Rational Optimist.*

94 **The Breakthrough Institute,** a US think tank, offers a good dose of what I call 'techno-optimism'. It is worth browsing their recent publications at http://thebreakthrough.org

95 The movement known as **Deep Ecology,** founded by the Norwegian philosopher Arne Naess, would reject what it would call the 'shallow ecology' of the thinking in books like this one. To people like Naess, humankind's massive disruption of natural systems has to be intensively questioned. The best place to find more is at The Foundation for Deep Ecology. See http://www.deepecology.org/movement.htm

96 The other main current in ecology that I have not covered properly in the book is what is known as 'steady state economics'. See http://steadystate.org for analysis of why economic growth does not deliver increases in human welfare and destroys the environment. I recognize the strength of this position but my view is that growth can help us achieve sustainability.

97 Europe may be behind the US on these issues. Apart from Tim Jackson's book *Prosperity Without Growth*, I have struggled to find writings that argue rigorously that sustainability requires us **to rein back growth.** In the UK, the New Economics Foundation (nef) seems to be the only think tank working largely in this area. I recommend a collection of short essays, co-edited by one of the nef's key figures. See *Do Good Lives Have to Cost the Earth?* (Constable, 2008).

98 My views on the impact of economic growth on society's ability to respond positively to challenges were partly formed by reading **Ben Friedman's** *The Moral Consequences of Economic Growth* (Vantage, 2005). Friedman's contention is that periods of prosperity are strongly correlated with improvements in society's values – its tolerance, support for the less privileged and interest in remaining inquiring. Not only do I believe that GDP growth is good for *technical* innovation, I believe it is probably important for developing a country's ability to develop *socially*, becoming more flexible and intellectually creative. We need these skills for sustainability. Finally, richer societies have more money to spend on the scientists and engineers needed to solve our terrifying problems.

99 We live today in **the cleanest, most disease-free and most resilient society in the Earth's history.** Our ability to dispose of human waste means that we probably live in the least polluted environment as well. In rich countries

(I'm not speaking about the developing world here) the areas of localized severe pollution, including poor air in city centres, need further work. But, in general, in the case of every type of pollution except greenhouse gases, we are either making progress or have a realistic scheme for improving things in the future.

100 The moral high ground on sustainability tends to be held by people who instinctively reject both the engineer's view of the world ('we can solve this problem by using technology intelligently') and the economist's ('the price mechanism will cure this problem without us doing much'). We need the ecologists, the engineers and the economists to be working in concert instead of shouting at each other. There is no single right answer.

Acknowledgements

The author and publishers would like to thank the following for their permission to reproduce photos in this book. **Chapter 1, Figure 1.1:** © Kbh3rd/http://en.wikipedia.org/wiki/File:Ogallala_saturated_thickness_1997-sattk97-v2.svg/http://creativecommons.org/licenses/by-sa/3.0/deed.en; **Chapter 8, Figure 8.1:** © NASA Earth Observatory; © Hypermania/Alamy.

Index

Annja examine
discovered

"What is that?" the leader of the would-be robbers asked. "Is that gold?"

Without a word, Annja got up and walked to the pile of driftwood the dig crew had gathered. She took out a three-foot piece that wasn't quite as big around as her wrist. She walked to the bound leader and twisted his arm, forcing him to stand. She ran the stick between his elbows and pulled his arms back. Before he knew what was happening, Annja tripped him and sent him face first onto the ground so that he was turned away from her working area.

She glared at the other prisoners. "Anyone else want to spend the night on his stomach?"

The three men looked away.

Annja felt bad about the rough way she'd treated the man. But they were out in the jungle, away from the civilized world, and the man had found a way to terrorize the dig team even while restrained. She didn't owe him any kindness.

She quashed her guilt and returned to work.

Titles in this series:

ROGUE Angel

Alex Archer

SERPENT'S KISS

A GOLD EAGLE BOOK FROM
WORLDWIDE®

TORONTO • NEW YORK • LONDON
AMSTERDAM • PARIS • SYDNEY • HAMBURG
STOCKHOLM • ATHENS • TOKYO • MILAN
MADRID • WARSAW • BUDAPEST • AUCKLAND

First edition January 2008

ISBN-13: 978-0-373-62128-6
ISBN-10:   0-373-62128-0

SERPENT'S KISS

Special thanks and acknowledgment to
Mel Odom for his contribution to this work.

Printed in U.S.A.

The
LEGEND

...THE ENGLISH COMMANDER TOOK
JOAN'S SWORD AND RAISED IT HIGH.

The broadsword, plain and unadorned,
gleamed in the firelight. He put the tip against
the ground and his foot at the center of the blade.
The broadsword shattered, fragments falling
into the mud. The crowd surged forward,
peasant and soldier, and snatched the shards
from the trampled mud. The commander tossed
the hilt deep into the crowd.
Smoke almost obscured Joan, but she continued
praying till the end, until finally the flames climbed
her body and she sagged against the restraints.

Joan of Arc died that fateful day in France,
but her legend and sword are reborn....

PROLOGUE

Kaveripattinam, India
509 B.C.

Sahadeva held the porcelain plate and pretended to examine it as he scanned the marketplace behind him. His heart, already beating quickly, nearly exploded when he saw their pursuers.

"They're still there, Sahadeva."

Jyotsna's whisper barely reached Sahadeva's ears. He felt her trembling at his side. The marketplace offered untold terrors for her. She'd never been in a place so big or so filled with people. Knowing that they had enemies nearby only made things worse.

Carefully, so he wouldn't incur the ire of the merchant, Sahadeva replaced the plate on the stack. The merchant started haggling, but the attempt lacked passion. Sahadeva's worn and dirty clothing warned all of the shopkeepers and traders that he lacked money.

After thanking the man and praising his goods, Sahadeva took Jyotsna's hand and led her toward the alley at the shop's side. He touched the curved knife in the sash at his waist. He'd never killed a man before. He didn't even like slaughtering the goats to put on the family table.

But he knew he would kill the men who pursued them in order to protect Jyotsna.

She looked like a child next to him. The top of her head barely came to his shoulder. Even draped in a loose dark-blue sari anyone could see that she had a woman's curves. Sahadeva worried her beauty might bring trouble to them in the city. A plain *dupatta* covered her head and held her thick black hair out of her face.

Sahadeva was young and slim. All of his life he'd been a goatherd. Nearly a year ago, when he'd turned seventeen, he'd run away from home to join a group of young men who'd decided to take a boat up the Vaigai River. Legends of gold and silver, of lost fortunes and fantastic monsters, had beckoned.

When he'd left, Sahadeva had known his father would be angry with him and his mother would be disappointed. Three days into the journey, he'd been frightened and doubtful despite the stories of adventure. Nine days later, just when they'd been about to exhaust their stores and forced to return home empty-handed, he'd seen Jyotsna and fallen in love with her.

She'd wanted to see the big world he described. Her father had denied that to her as he had denied it to all his people. Only the warriors had ventured outside the cave city to get food. Occasionally they brought brides and grooms back into their secret village.

Those brides and grooms, he'd discovered, had only been allowed to live there for a short time. Outsiders were put to death once the children were planted. Sahadeva had seen monstrous things among Jyotsna's people. There was no sign of anyone who had come from outside their enclave to live among them.

Jyotsna had captured Sahadeva's heart. And she had been equally drawn to him. Unable to bear the thought of his death, she had warned him of the coming assassinations. Sahadeva talked her into running away with him, and they fled.

Now all of his friends were dead. Jyotsna's father's warriors had killed them mercilessly. Only luck and his knowledge of the terrain along the Vaigai had prevented Sahadeva and Jyotsna from getting overtaken.

But those pursuers were here now. Even Kaveripattinam, as large as it was and open to trade around the world, wasn't enough to hide them.

Sahadeva strode briskly through the marketplace, past the shops and hawkers, through the maze of goods and buyers, until he reached the alley. Voices, whistles, bells and animal bleats sounded all around him.

The buildings flanking the alley blocked some of the heat of the midmorning sun in the narrow expanse. By noon Sahadeva knew the stones beneath his callused feet would be blistering.

At the other end of the alley, he a saw the harbor spread out before him. Tall Roman galleys sat in the ocean. And there were more vessels from other countries.

Since he'd been a boy and his father had first allowed him to help drive goats to market, Sahadeva had loved the sea. The sailors with their stories of foreign lands and exotic sights had filled his head. When he'd talked to his father about such things, his father had told him to quit wasting his time dreaming. He'd said a goatherd would never have enough money to buy a ship, and taking passage on one as a sailor was nothing short of slavery.

Things change, Father, Sahadeva thought grimly. He approached a man arranging a cart filled with woven baskets. "Sir," he said. "I'm looking for Harshad the jeweler."

The man stroked his fingers in his long beard then pointed. "Harshad's shop is in the next street. On the right."

Sahadeva thanked him and got moving again. The crowd was thinner. He didn't think the men who pursued them would do anything here, but there were no guarantees. They were desperate men. He'd taken more than Jyotsna when he'd left their city.

A BURLY MAN STOOD guard at the jeweler's door. He looked half-asleep, but the sword through his sash was sharp and nicked from use. Scars showed on his thick arms.

When he started to enter, the guard put his big hand in the middle of Sahadeva's chest and stopped him. "There's no begging allowed in this shop."

Despite his fear and the urgency that pressed him, Sahadeva's pride burned. "I'm no beggar." His hand dropped to his knife.

The guard smiled. "You're wearing a beggar's rags, boy. And I wouldn't pull out that knife. Unless you're ready to die."

Sahadeva swallowed hard and felt his face burn with shame. "I've got business with Harshad." He reached inside his shirt and took out a small oilskin pouch. Another oilskin bag was hidden inside the pack he carried, but thieves wouldn't have wanted it. Still, he never left it unattended. "I have merchandise for sale."

Sunlight glinted off the gold and gems inside the bag.

Jyotsna's fingernails bit into Sahadeva's arm. "What have you done?"

Sahadeva looked into her dark eyes. "I did what I had to so that we could be together."

Tears glinted in her gaze and she looked away from him.

Sahadeva felt torn. He didn't have time to explain. Jyotsna had always lived within her father's village. She had no idea what the real world was like or what it took to live in it.

"Send the boy in here," a man's voice called from within the shop.

Reluctantly, the guard stepped aside.

Sahadeva moved forward. He had to pull on Jyotsna's arm twice to get her to follow him.

Inside, the shop was small and heavily scented with incense. A thin man with graying hair and beard stood behind a counter. He wore a white tunic. Earrings, rings, necklaces, hair bands and gold-and-silver bangles hung from pegs on the wall behind him. Jewels sparkled in settings in some of them. Harshad smiled. "Welcome. What may I do for you?"

Sahadeva freed his arm from Jyotsna. He placed the oilskin pouch on the counter. "I want to trade these for gold coins."

The jeweler spread the bag's contents across the counter. Five rings, two bracelets and a loose collection of gems spread between them. Harshad looked at the jewelry with marked interest. "These are of very unusual design. Where did you get them?"

"I found them," Sahadeva replied. "They were in the Vaigai River."

The jeweler looked up. "Where in the Vaigai?"

Sahadeva shook his head. "There isn't any more there."

"Maybe you just didn't look closely enough."

"Then I'll go back and look again."

Harshad frowned. "You've been most fortunate, it seems."

"How much?" Sahadeva asked.

"Are you in a hurry?"

"No," Sahadeva lied. He'd learned at his father's knee never to show impatience during a trade. A hasty man often got the worst of a bargain. A needy man fared even worse.

But what about a man who fears for his life? Sahadeva wondered.

"I could look at these and give you an offer tomorrow," Harshad suggested.

"By that time I could get offers from other jewelers," Sahadeva countered. "I was told I could get a fair price from you."

"Wandering around the city could be dangerous," Harshad said.

Sahadeva started gathering his treasure. "The ships are in. I want passage on one of them. Perhaps I can strike a deal with a captain who would trust his instinct for jewelry."

"Wait," Harshad said. He sighed. "I'm only going to do this because you look like a good boy. Although some might question if you really got these things from the river."

Sahadeva held the bag tightly. He'd come to Harshad because he'd heard the man didn't ask too many questions.

"Come with me." Harshad gestured to a doorway draped by curtains. He stepped through them and waved again.

Sahadeva and Jyotsna followed him.

"Just you," the jeweler said. "Back here I only deal one-on-one."

Sahadeva hesitated, then turned to Jyotsna. "Stay here."

She pulled on his arm. "Don't leave me."

"It'll only be for a moment. You'll be safe." Sahadeva gently pulled her hands from his arm. Doing so almost broke his heart because her fear showed in her liquid gaze. "I'll be right back. I promise."

Jyotsna wrapped her arms around herself. As she stood there, she looked incredibly small.

Sahadeva made himself turn and follow the jeweler to the back room of the shop.

"Sit, sit." Harshad gestured toward a chair on the other side of a small wooden table in the back room.

Sahadeva waved away the thick white smoke given off

by the incense. Coils of the fragrant paste burned in every corner of the room. He sat at the table. The smoke made it hard to breathe and he immediately felt light-headed.

Harshad clapped his hands. Immediately an old woman appeared through another door and delivered a tea service. She poured *cocum* squash into tall glasses of water, then left without a word.

Sahadeva's taste buds flooded at the drink's scent. He and Jyotsna had subsisted on bread, goat's cheese and water. *Cocum* squash was only available in April and May. He'd almost missed the season entirely. He picked up the glass and felt the chill.

"Let me again examine what you have," Harshad said. He smiled once more.

Sahadeva saw the anticipation in the man's face. Harshad clearly wanted the jewelry and gems. Slowly, Sahadeva emptied the pouch onto the table. The heavy gold smacked into the wood. The sound echoed strangely in Sahadeva's ears.

"You found these in the Vaigai River, you say?" Harshad examined one of the rings.

"Yes," Sahadeva said. He sipped his drink. The flavor was strong and cool.

"You're lucky. Many men have searched that river for treasure," Harshad said.

"I know."

"Some soothsayers still insist there is a secret city with impossible wealth located there."

Sahadeva's heart thudded and his head swelled from the pressure. "I wouldn't know about that," he said.

"It is supposed to be a city of *naga*s," Harshad said

as he moved on to examine a bracelet. "Half men, half snakes. Have you ever seen such a thing?"

"No." But Sahadeva knew well the old stories and legends told of such things.

"They lived on an island, it's said. Then the monsoon season brought a wave that broke their island and drove them inland. They tried to live on the mainland, but they worshiped snakes and practiced bloodthirsty rituals. No one would suffer them to live there. So they fled upriver."

Sahadeva listened without comment. He had to force himself to breathe. He wanted out of the room. Anxiety crawled over him at the thought of leaving Jyotsna with the burly guard. It was worse thinking about her father's warriors lurking in the street.

"Do you think these things came from that city?" Harshad asked.

Sahadeva's heartbeat became thunder in his ears. He was certain the jeweler could hear it. "No," he lied.

"Why not?" Harshad asked.

"No one has ever proved that city ever existed," Sahadeva said. No one had ever found the tributary Sahadeva and his friends had found, either. It went underground for a time, and if Pramath hadn't gone hunting that morning they might never have found it, he thought.

"Still," Harshad mused, "there is usually some kernel of truth in those old legends."

Sahadeva said nothing. He pulled at his collar in an effort to get more air. Heat flushed his face. He forced air into his lungs.

"I've even been told that the things that have been found from the *naga* city are cursed," Harshad said.

"Cursed?" Sahadeva's mind tried to grasp the word but it slipped away.

"I've been told," Harshad said in a quiet voice, "that the *naga* spirits follow anything that was taken from their city. They find them and bring them back after killing those who stole them. Do you believe in curses?"

Sahadeva thought about that for a moment while he finished the rest of his drink. He'd never actually seen a curse in effect, but he'd heard stories about them all of his life. "I don't know," he finally said.

"Well, it's better to keep an open mind, perhaps. When you've lived as many years as I have, you'll learn the wise men don't have all the answers." Harshad pushed the jewelry and gems to the center of the table. "Now we must discuss what these are worth to you."

For the next few minutes, they haggled over the price. Sahadeva knew not to take the first offer. Only a fool and an amateur took the first offer. His father had taught him that, as well.

Finally, they agreed upon an amount. Sahadeva didn't know if it was fair, but it was more than he'd been hoping to get for the pieces. He was certain Harshad thought he'd gotten the better of the bargain.

Sahadeva wanted only enough to arrange passage on one of the ships in the harbor. He knew he and Jyotsna would have to start over somewhere new. Perhaps Greece or Rome would be a good choice. He might even like to see Egypt. Those countries accepted foreigners.

Besides, he hadn't shown Harshad the full treasure they'd escaped with.

"I must tell you one thing," Harshad said at the end

of the negotiations. "If these things are indeed cursed, I expect you to take them back. Is this understood?"

Sahadeva readily agreed. He didn't believe in the curse. Even so, he would be long gone in just a matter of hours if he could find a ship putting out to sea in that time.

"I will return with your gold." Harshad got up and left the room. He left the jewelry and gems sitting on the table.

Sahadeva felt his head grow heavier. When he turned to look at the window high on the wall, his senses whirled. He realized the colors seemed brighter than normal, and the sounds coming from outside were leaden and muffled.

Something was wrong.

He tried to stand but his legs were almost too weak to hold his weight. He gasped for air and choked on the thick incense smoke. He tried to sweep the jewelry and gems into the pouch again, but only succeeded in scattering them across the table and the floor.

A cloud of smoke suddenly burst inside the room. A loud hiss accompanied it.

Startled, Sahadeva stumbled back against the wall. The acrid smoke burned his nose and throat when he inhaled it. Incredulous, he watched as a figure took shape.

The head and shoulders of a beautiful woman appeared first. Jeweled combs pinned her thick black hair atop her head. Her garments barely covered her modesty, like the garments Jyotsna's people wore. She stood high-breasted and proud. She peered at him with the slit-irised eyes of a cat. Crimson lips parted to reveal sharp teeth. Her forked tongue slithered out to test the air.

As she moved toward Sahadeva, she rocked from

side to side. Her lower half was hidden from sight by the smoke for a moment. When he saw the serpentine body that began at her waist, he tried to scream but there wasn't enough air in the room.

From her midriff down, the woman was a snake. Glittering blue-green scales twisted as she moved. Black-and-red scales created a hard-edged pattern. In the next instant, she lunged at him and her fangs pierced his throat.

SAHADEVA WOKE to a pounding pain in his head. Blood roared in his ears. He felt dizzy, as if the world were shifting beneath him. He opened his eyes and discovered the reason for the movement.

He was in a ship's hold. The light from a candle on a mounted sconce barely penetrated the gloom. He lay in the middle of a pool of vomit that he realized was his own. It had smeared on his clothing and made the fabric stiff. Iron manacles bound his legs to a ring set in the floor.

Where is Jyotsna? The question drove him to his feet in spite of the pain and sickness coiled in his belly. He immediately threw up again.

"Easy, now," someone said from the darkness.

The ship tossed and turned. Timbers creaked in protest. The floor tilted so much for a moment that Sahadeva feared they were going to turn over.

Sahadeva tracked the voice and saw a man in his middle years sitting hunched against the wall. Nine others sat with him.

"Who are you?" Sahadeva asked. "What is this ship?"

"I'm a slave," the man answered. "Like you. My name is Oorjit."

"I'm not a slave," Sahadeva objected.

"You lie in your own filth aboard a ship that you didn't book passage on," another man said. "You're a slave. When the captain has outrun this storm, they'll bring us up and start making sailors of us."

"I'm afraid what he says is true," Oorjit said. "All of us were taken in Kaveripattinam. The ships' captains do this when they need crew and no one is willing to sign on. Lives are cheap in the city. Doubtless you were sold into captivity by someone who profited in the loss of your freedom."

Sahadeva slumped in disbelief. His first thoughts were of Jyotsna. He'd brought her to the city and told her he could take care of her. He wept when he thought of the horrors he had doubtlessly left her to face.

The ship continued to roll. The movement grew more violent. Water sloshed around Sahadeva's ankles and he thought it was growing deeper.

"Are you a soothsayer?" Oorjit asked.

Sahadeva looked at the man.

"The book you carried." Oorjit tossed over Sahadeva's battered travel pack. "I thought if you could read you were a soothsayer."

Sahadeva couldn't believe the pack had been left with him. He doubted the other gems and jewelry remained. He searched the bag and found he was correct. But the book lay there.

"Doubtless they couldn't find anything in it worth stealing," Oorjit stated. "Slavers aren't readers."

Sahadeva picked up the book and examined it. He'd stolen it from the treasure room in Jyotsna's village. He knew wise men and kings often paid handsomely for such things.

It was a thick rectangle covered in some kind of hide. Sahadeva thought it was snakeskin, but he wasn't sure. He felt the binding and made sure the thing he'd hidden there remained. He'd kept that from Harshad. Sahadeva had intended to use it to buy a business for himself whenever they got to where they were going.

Even though it remained, Sahadeva knew that the future he'd planned was gone. He replaced the book in its protective oilskin and shoved it under his shirt.

"Are you a wise man?" Oorjit asked.

"No," Sahadeva answered.

"Pity," the man said. "I think it would be good to have the ears of the gods in this storm."

Sahadeva didn't know how much time passed in the hold. The candle flame wavered as the ship heaved and rolled. Several times he felt as though the sea had pitched them into the air. As the sick fear grew inside him, he knew what was going to happen. It was only a matter of time.

Still, when it did occur he wasn't prepared. The ship capsized. Water rushed into the hold. With his legs chained, he had no chance. Despite his best efforts and his most impassioned pleas, the cruel, uncaring sea swallowed him.

1

Annja Creed stood in a twelve-foot-deep sacrificial pit beneath a gathering storm. The storm, according to the weather reports, was hours away but promised to be severe. From the look of the skeletons on the floor of the pit and embedded in the walls, hundreds of years had passed since the last sacrifice.

The passage of time hadn't made the discovery any less chilling. Even with her experience as an archeologist—and the recent exposures to sudden death that she thought were incited by the mystic sword she'd inherited—she still had to make the conscious mental shift from personal empathy to scientific detachment.

"Are those *human* bones?"

Annja glanced up and saw Jason Kim standing near the edge of the pit above her. Jason was a UCLA graduate student who'd won a place on Professor Rai's dig along the southern coast of India.

Jason was barely over five and a half feet tall and slender as a reed. His long black hair blew in the strong wind summoned by the storm gathering somewhere over the Indian Ocean. Thick glasses covered his eyes, which were bloodshot from staying up too late playing PSP games in his tent. He came from a traditional Chinese family that hated the way he'd so easily acquired American ways. He wore a concert T-shirt and jean shorts. A tuft of whiskers barely smudged his pointed chin.

"They're human bones," Annja answered.

"You think they're sacrifice victims?" Jason's immediate interest sounded bloodthirsty, but Annja knew it was only curiosity.

"I do." Annja knelt and scooped one of the skulls from the loose soil at the bottom of the pit. She indicated the uneven cut through the spine at the base of the skull. "Followers of Shakti favored decapitation."

"Cool. Can I see that?" Jason held his hands out.

Annja only thought for a moment that the skull had once housed a human being. The truth was, in her work, the body left behind was as much a temporary shelter as the homes she unearthed and studied.

Jason's field of study was forensic anthropology. His work primarily included what was left of a body. If anyone at the dig could identify the tool marks on the skeleton, it was Jason.

Annja tossed the skull up to him.

Jason caught the skull in both hands. It didn't bother him that it was so fresh from the grave. His smile went from ear to ear. He rotated the skull in his fingers. "This is the bomb, Annja."

"Glad you like it."

"Think they'll let me keep one?" he asked.

Part of Annja couldn't believe he'd asked the question. The other part of her couldn't believe she hadn't expected it.

"Definitely not," she answered.

"Too bad. Put a small, battery-operated red light inside and this thing would be totally rad. I could even have a friend of mine majoring in dentistry whip up some caps for the incisors. I'd be the first guy to have a genuine vampire skull."

"Except for the *genuine* part. And you'd have to explain why the skull doesn't turn to dust in sunlight," Annja said.

"Not all vampires turn to dust. You should know that," he replied.

"Vampires aren't a big part of archaeology." Annja turned her attention back to the other bones. She didn't think she was going to learn a lot from the pit, but there were always surprises.

"I didn't mean from archaeology," Jason persisted. "I mean from your show."

Annja sighed. No matter where she went, except for highly academic circles, she invariably ended up being known more for her work on *Chasing History's Monsters* than anything else. The syndicated television show had gone international almost overnight, and was continuing to do well in the ratings.

Scenes from stories she'd done for the show had ended up on magazine covers, on YouTube and other television shows. Her producer, Doug Morrell, never missed an opportunity to promote the show.

"You ever watch the show?" Annja looked up at Jason and couldn't believe she was having the conversation with him.

"Sure. The frat guys go nuts for it. So do the sororities. I mean, DVR means never having to miss a television show again."

Terrific, Annja thought.

"Kind of divided loyalties, though," Jason said. "The sororities watch you." He shrugged. "Well, most of them do. The frat guys like to watch the show for Kristie."

Okay, I really didn't need to hear that, Annja thought.

Kristie Chatham, the other hostess of *Chasing History's Monsters,* wasn't a rival. At least, Annja didn't see Kristie as such. Kristie wasn't an archaeologist and didn't care about history. Or even about getting the facts straight.

When Kristie put her stories together, they were strictly for shock value. As a result, Kristie's stories tended to center on werewolves, vampires, serial killers and escaped lab experiments.

"You can't go into a frat house without finding her new poster," Jason went on.

"That's good to know," Annja said, then realized that maybe she'd responded a little more coldly than she'd intended.

"Hey." Jason held his hands up in defense and almost dropped his newly acquired skull. He bobbled it and managed to hang on to it. "I didn't mean anything by that."

"No problem," Annja said.

"I don't know why you don't do a poster," Jason said. "You're beautiful."

Maybe if the comment hadn't come from a geeky male in his early twenties who was five years her junior and had a skull under his arm, if she hadn't been covered in dirt from the sacrificial pit and perspiring heavily from the gathering storm's humidity, Annja might have taken solace in that compliment.

Dressed in khaki cargo shorts, hiking boots and a gray pullover, she stood five feet ten inches tall and had a full figure instead of the anorexic look favored by so many modeling agencies. She wore her chestnut-brown hair pulled back under a New York Yankees baseball cap. Her startling amber-green eyes never failed to capture attention.

"I don't do a poster because I don't want to end up on the walls of frat houses," Annja said.

"Or ceilings," Jason said. "A lot of guys put Kristie's posters on the ceiling."

Lightning flashed in the leaden sky and highlighted the dark clouds. Shortly afterward, peals of thunder slammed into the beach.

Jason looked up. "Man, this is gonna suck. I hate getting wet."

"That's part of the job," Annja told him. "The other part is being too hot, too tired, too claustrophobic and a thousand other discomforts I could name."

"I know. But that's only if I stay with fieldwork. I'd rather get a job at a museum. Or in a crime lab working forensics."

Annja was disappointed to hear that. Jason Kim was a good student. He was going to be a good forensic anthropologist. She couldn't understand why anyone would choose to stay indoors in a job that could take them anywhere in the world.

Lightning flashed again. The wind shifted and swept into the pit where Annja stood. The humidity increased and felt like an impossible burden.

"I'm gonna go clean this up," Jason said. "Maybe after we batten down the hatches, you can tell me more about who Shakti was."

Annja nodded and turned her attention back to the burial site. The storm was coming and there was no time to waste.

WITH CAREFUL DELIBERATION, Annja checked the scale representation of the burial pit she'd drawn. So far everything was going easily, but she suspected it was the calm before the storm.

The drawing looked good. She'd also backed up the sketch with several captured digital images using her camera. In the old days, archaeologists only had a pad and paper to record data and findings. She liked working that way. It felt as if it kept her in touch with the roots of her chosen field.

She stared at the body she'd exhumed. From the flared hips, she felt certain that the bones had been a woman. She resolved to have Jason make the final call on that, though.

Lightning flickered and thunder pealed almost immediately after. The storm was drawing closer.

"Annja."

Glancing up, Annja spotted the elfin figure of Professor Lochata Rai, the dig's supervisor. Lochata was only five feet tall and weighed about ninety pounds. She was in her early sixties, but still spry and driven. She wore khakis and looked ready for a trek across the Gobi Desert.

"It is time for you to rise up out of there. The rain is coming," the professor said.

Annja looked past the woman at the scudding clouds that filled the sky. Irritation flared through her at the time she was losing.

"We must cover this excavation pit," Lochata said. "Perhaps it will not rain too hard and we won't lose anything."

"I know. This really stinks because we just got down far enough to take a good look at what's here," Annja said.

Lochata squatted at the edge of the pit. She held her pith helmet in her tiny hands over her knees. "You're too impatient. You have your whole life ahead of you, and history isn't going anywhere. This site will be here tomorrow."

"I keep telling myself that. But I also keep telling myself that once I finish this I can move on to something else." Annja stowed her gear in her backpack.

Lochata shook her head. "You expect to find something exciting and different?"

"I *hope* to." Annja pulled her backpack over her shoulder and climbed the narrow wooden ladder out of the pit. "I always hope to."

"I do not." Lochata offered her hand as Annja neared the top. "Finding something you did not expect means you didn't do your research properly. It also means extra work and possibly having to call someone else in to verify what you have found."

Annja understood that, but she also liked the idea of

the new, the undiscovered and the unexpected. Lately, her life had been filled with that. She thought she was growing addicted to it.

Once on the ground outside the pit, Annja stood with her arms out from her sides as if she were going to take flight. The wind blew almost hard enough to move her. Perspiration had soaked her clothing.

"Drink." Lochata held out a water bottle and smiled. "Hydrate or die."

Annja smiled back and accepted the water. The rule was a basic one for anyone who challenged the elements. She opened the bottle and drank deeply.

The dig site was in the jungle fringe that bordered the Indian Ocean. Kanyakumari lay as far south on the Indian continent as a person could go. They were forty miles west of there on a cliff twenty-seven feet above sea level. The ocean stretched to the south under the whirling storm clouds. Whitecaps broke the dark-blue surface.

"What are you thinking?" Lochata asked.

Annja grinned self-consciously. She didn't like to get caught daydreaming. The nuns who'd raised her in a New Orleans orphanage had worked hard to train that distraction out of her. It hadn't worked.

"I was just thinking about how many ships have been through those waters," Annja admitted.

"Ah, yes." Lochata's eyes glittered. "The Romans, the Egyptians, ships from China's Ming Dynasty."

"Vasco da Gama was the first European to sail the Indian Ocean," Annja said. "He was looking for a trade route around the Cape of Good Hope in Africa."

"The British took over after that," Lochata said.

"They brought their ships loaded with cannons and fought wars to control the area. The Dutch East India Company fought trade wars with the French and others."

"It isn't just history out there," Annja said wistfully. "Sinbad sailed those seas, as well."

Lochata laughed. "My, my. Bringing up fictional characters. You are the romantic, aren't you?"

"I try not to be," Annja said. "But if you think past this moment, if you see into the past, it's hard not to be." She paused as she watched the storm-tossed waves. "A lot of those ships didn't make it across the ocean. Storms took them, they were lost in sea battles and sometimes ships just went down."

"Or perhaps sea monsters got them," Lochata said laughing.

"I don't believe in sea monsters." Over the past few years, Annja had learned to believe in a lot of things, but she hadn't yet crossed paths with a sea monster.

"Perhaps not," Lochata said. "But the sea is a cruel mistress. She takes what she wants. She breaks the weak and the foolish. And she gives back only what she wants to."

Surprised, Annja looked at the older woman. "I didn't know you were a poet."

Lochata smiled and shook her head. "Not me. My husband. He's been in the merchant marine since he was a boy." Concern touched her dark features. "I worry about him a lot these days, but he won't give up the sea. A few years ago, things were not so dangerous out on the water. There are too many pirates out here now. They take what they want, and they kill and destroy."

Annja didn't say anything. She knew the professor was right. Before leaving her home in Brooklyn, she'd researched the area's past and present. The Indian Ocean pirates plying their trade were every bit as dangerous as the Shakti followers who had sacrificed so many innocents to their cruel goddess.

"We need to get everyone together," Lochata said as she gazed at the storm clouds above them. "I think this is going to be a bad one."

The wind picked up and rattled through the nearby trees.

"I thought monsoon season was over," Annja said.

Worry tightened the lines of the older woman's face. "It should be. I think this is something else." She looked at Annja and smiled. "You don't believe in curses, do you?"

Considering everything she'd been through since she'd found the final piece of Joan of Arc's legendary sword, Annja wasn't sure how to answer the question. Receiving the sword had changed her perspective on a lot of things.

"Not really," Annja finally said.

"Neither do I," Lochata agreed. "But we've been disturbing the final resting places of the dead. That's taboo in almost every culture."

2

"You're in India?"

Annja held the satellite phone to her ear and strained to hear. "Yes, Doug. India."

Doug Morrell, her producer, was twenty-two and excitable. He asked questions, but he only heard what he wanted to hear.

"India, as in half-a-world-away India?" Doug asked.

"Yes." Annja stood in the main tent and gazed out at the jungle. She and Professor Rai had gathered the dig team together.

Normally a break in the dull routine of digging would have been welcome. However, now they were all trapped in the leaking tent and hoping it would stay erect against the gale-force winds.

A torrential downpour slammed the surrounding jungle and reduced visibility to little more than a few yards. Beyond that everything turned gray and disappeared in the dark night. Annja could barely hear Doug over the crackle caused by lightning strikes and the heavy rain pounding against the canvas.

"What are you doing there?" Doug asked.

"I signed on to do a dig with Professor Lochata Rai."

"Uh-huh. So what's he digging up?"

"She."

"Okay. What's *she* digging up?"

"Professor Rai got permission from the Archaeological Survey of India to look for a Shakti sacrifice site."

"Did you just say *sacrifice?*" Doug's voice rose.

"I did." Annja regretted telling him that detail at once. If she hadn't been distracted by the building storm she wouldn't have.

"*Human* sacrifices?" Doug asked.

"And animal." Annja heard a keyboard clatter to life.

"Are you digging up human or animal bones?"

"Today we uncovered a pit containing several human skeletons."

"Bodacious." Doug's excitement grew. "Always interested in pieces on human sacrifice. Who did you say was doing the sacrifices?"

"Followers of Shakti." Annja spelled it out for him. She glanced back into the tent and saw the dig crew seated around long folding tables on a collection of lawn chairs.

Everyone on the crew was young. Most workers on archaeological excavations were interns or students. Generally there was barely enough money to fund a team with provisions, much less to make a profit. They sat playing board games, reading or telling stories. None of them acted like the storm worried them, but Annja knew they were concerned.

She was concerned.

"Shakti," Doug said. "Consort of Shiva."

"That's her." Annja sipped green tea from a bottle. It was one of her few extravagances for the dig. "That's not something you would know. You're looking on the Internet, aren't you?"

"You gotta love Wikipedia," Doug said.

Annja had written or corrected more than a few entries on subjects on the site.

"Wasn't Shiva the god of death or something?" Doug asked.

Annja really didn't want to get into a lesson on Hinduism. That would be a long discussion and Doug would only hear what he wanted.

"Yes," she replied. It was the simplest answer. Annja knew, as with all Hindu gods, Shiva was much more than one thing.

"This human-sacrifice thing has potential. We haven't done a piece on a god of death in months," Doug said.

"I'm not doing a story," Annja said. "I'm here to work a dig."

"I know, I know. I was just wondering if there was a way we could get a twofer."

"I'm not interested in a twofer. I came out here to work."

"Hey, don't bite the hand that feeds you."

Annja swallowed a sharp retort. She couldn't complain about the television show. *Chasing History's Monsters* had been good to her. Real archaeology didn't pay a lot. To be part of Lochata's dig Annja had

had to pay her own way over. The community meals were free and the cot was a loaner. She had to buy her own bottled green tea.

The television show offered the glamour and glory. It also came with a paycheck that enabled her to do things like this dig.

"Okay." Annja stared out at the dark sky. She couldn't see the edge of the cliff. The crash of the surf against the rocks below remained audible.

"Okay what?" Doug asked.

"What do you have in mind?"

"Annja Creed stalks mysterious cult that carries out human sacrifices."

"This particular cult's been dead for hundreds of years. Probably more than a thousand."

"They gotta have descendants, right?" Doug asked.

"Possibly, but I wouldn't know how to get in contact with—"

"I'm looking at a news story that says these Shakti cultists have been up to their old tricks in different parts of India."

"Old tricks?" Annja asked.

"Creative license on my part," Doug said. "Makes them sound more devious and threatening. Ups their coolness quotient, trust me. Anyway, there are Shakti cultists springing up. No human sacrifices have been found yet, but that may be because they've hidden the bodies. Or buried them."

Annja could tell Doug was selling himself on the idea.

"Maybe you could take some footage of the local jungle as you make your way through a forgotten trail."

"If it was a forgotten trail," Annja said, "I wouldn't know about it."

"Of course you would. You're a world-famous archaeologist."

Annja smiled a little at that. If Doug hadn't been trying so hard to flatter her, she might have enjoyed his efforts. But she'd known him long enough to be aware that he seldom did anything without an ulterior motive.

"How much longer are you going to be there?" Doug asked.

"A few more weeks."

"See? You can work in a piece on human sacrifices," he said.

"I'm busy. When you work a dig, you're putting in eighteen- to twenty-hour days."

"Don't you have a day off?"

"When I do, I like to have it as a day off." So far there hadn't been one of those. Annja watched one of the students run back through the jungle from the cliff area. The young woman's boots splashed across the drenched ground. Panic pulled her face tight. She was one of Professor Rai's students and knew the area well. If she was frightened, there had to be a reason.

"Doug," Annja interrupted as he launched into a guilt-inspiring speech, "I'm going to have to call you back." She closed the phone and put it into her pocket. She knew Doug hated being hung up on and wasn't surprised when he called right back. Annja ignored the ring tone and lunged out into the driving rain.

Lochata ran out to meet the student and reached her before Annja. The older woman grabbed the younger one by the shoulders and forced her to calm enough to talk. They spoke rapidly in their native tongue, and Annja didn't understand a word. The student kept gesturing toward the cliff.

Her boots heavy with the mud that had collected on them, Annja joined the professor and student. Rivulets ran down the bill of Annja's baseball cap, and she was drenched at once. She reached into the otherwhere and felt the sword. The hilt felt familiar in her hand and she took comfort in it.

From the reddened state of the student's eyes, Annja knew she was crying. But the tears mixed in with the rain so quickly they disappeared at once.

"What's wrong?" Annja asked.

Lochata gathered the young woman into her embrace for a moment, then spoke soothingly to her and pushed her toward the main tent. Immediately the professor headed toward the cliff. "She says the sea has withdrawn," Lochata stated.

"Withdrawn?" Annja matched the older woman's stride.

"Receded."

"An outgoing tide will do that."

"She says this is more than just the tide." Lochata's face looked grave.

Annja studied the irregular line of broken rocks at the foot of the cliff. They had been at the dig site for five days. She'd walked out to the cliff on several occasions to take a break from digging through the hard-packed earth and stared out at the ocean.

She'd never seen the rocks or that much of the sea

bottom before. As she watched, the water seemed to draw back even more.

"The sea's never done that before," Annja said.

Lochata's face drained of color. She turned to face Annja. "Tsunami," she said, and the hammering thunder overhead almost swept her words away.

Fear shook Annja.

The horrifying images of the December 26, 2004, tsunami had shocked the world. And the devastating waves killed a quarter of a million people. She grabbed Lochata's arm. "Run!" She pushed the older woman into motion.

Despite her age, the archaeology professor proved fleet-footed. She ran through the dig site and avoided the pits the team had dug in their search for the sacrificial well.

Together, they ducked through the trees and scrambled through the bushes. Lochata stumbled and would have fallen twice, but Annja caught her and kept her vertical. Then, just when the tents became visible, the ground shook so hard that Lochata and Annja both lost their footing and went down.

Mud coated Annja's clothing and the right side of her face. She wiped it out of her right eye and tried to ignore the burning sensation it caused. She pushed herself up and hauled Lochata to her feet.

Unable to stop herself, Annja looked back toward the sea. In the distance, barely discernible through the haze of fog, a giant wall of water raced toward the coast.

For one frozen second, Annja stood locked in place. Even with everything she'd seen with Roux and Garin, she wasn't prepared for the tsunami wave. It was a huge, rolling curl of ocean that was closing on the shoreline quickly.

At first Annja had hoped that the cliff might be above the crest line of the tsunami. The wave that had struck the coastline in 2004 was reportedly 108 feet high. Annja couldn't tell how high the water was, but she could see it was higher than the cliff.

The site crew stared at the approaching wave in open-mouthed horror. Then the screaming started.

"Get into the trees!" Annja yelled. "Climb the trees!"

She didn't know if that would work, but there was no way they were going to outrun the wave. Climbing was the only option.

"The trees!" Annja yelled again.

Lochata gave more orders in her native tongue.

The dig crew started climbing trees as another tremor shook the ground. Annja had a quick image of the cliff shearing off and plunging into the sea with them on it. It was a terrifying thought.

She ducked into the small tent she'd been issued, grabbed her backpack and slid it over one shoulder. On her way to the nearest tree, she took a coil of rope with a grappling hook from the back of one of the four-wheel-drive vehicles they had to help with transportation.

Rope was always important on a dig, and she knew it would come in handy while they were in the trees. If nothing else, they could use it to lash their supplies together or as a safety line until the floodwaters subsided.

The ground rumbled again. The approaching wave drowned out all other sounds.

Nearby, one of the young men opened the door to one

of the SUVs and tried to clamber inside. Annja grabbed the young man's shoulder. She pulled him out onto the muddy ground harder than she'd intended. He hit the ground and rolled, but fear gave him springs and he bounced up at once.

"What are you doing?" the young man demanded. His name was Nigel. He was one of the Brits on the team. He'd been something of a troublemaker who didn't always pull his full shift and couldn't be counted on to be thorough. Many of the team had started resenting him.

"If you climb in that vehicle you're going to drown," Annja shouted over the growing roar.

"We can't stay here, you bloody cow!" Nigel started for the SUV again.

Annja moved into his path.

Nigel threw a vicious punch at Annja's head.

Annja shifted, dropped her hips to lower her center of gravity, blocked his punch with the back of her left wrist and turned it outside, away from her head. She responded with a jab and almost caught him full in the face with it before she opened her hand to slap him.

The young man went down again. This time he remained on his hands and knees for a moment while his senses whirled. He spit curses.

Ignoring the venom in his tone, Annja grabbed him by the shoulder and hustled him to the nearest tree large enough to support their weight. Four other site workers had already taken refuge among the thick branches.

"Higher!" Annja shouted.

The others clambered higher.

"Get up the tree, Nigel." Annja pressed him against the rough bark of the teak tree.

Teak grew freely in southern India, especially in the Tamil Nadu district. The trees towered over a hundred feet and provided plenty of branches for the climbers to use to haul themselves up.

Growling curses, Nigel climbed the tree. Annja followed just as the wave smashed into the cliff face hard enough to make the ground shiver. In the next instant, the wave rolled into Annja like a battering ram and knocked her into the tree trunk with enough force to stun her.

The rough bark smashed into the side of her face, and burning pain followed. She hung on to the tree out of desperation, but as she realized her grip would hold, the wave rolled over her. In the next second she was underwater and drowning.

3

Annja clung to the bark and hoped it didn't give way beneath her fingers. All she saw was swirling water. The floodwaters muted her hearing, but she heard her heart beating frantically. Stay calm, she told herself. You're going to be all right.

She knew from experience that she was prone to tell herself that lie every time things turned out badly.

She tilted her head back and looked up the tree. She couldn't tell how high the water went. She felt the tree quiver under the onslaught of the flood.

Staying underwater wasn't an option. Grimly, Annja slid her arms and legs up the trunk and felt the bark bite deeply into her bare flesh. She crept up slowly, only inches at a time.

Just as her lungs felt near to burst, her head broke the surface. She managed a quick breath and turned seaward. Another wave slapped her in the face and almost knocked her from her precarious hold. She caught a branch above her head and hauled herself out of the water.

Almost full dark had rolled in with the tsunami. Annja surveyed the trees for the survivors. The roar of the water made conversation almost impossible. But she heard her name.

"Annja!" A flashlight beam lanced through the darkness.

"Here!" Annja shouted.

The bright beam stung her eyes. She turned her head away. She sat on a branch eight feet above the water. The level didn't appear to be rising.

"They said you were underwater," Lochata shouted from the nearby tree.

"Not anymore." Judging herself to be at a safe height, Annja shrugged out of her backpack and checked it. It was constructed so the main cargo area, where she carried her notebook computer, her camera and her other electronic equipment, was waterproof. She'd carefully packed it when the storm approached. The only worry was that debris might have smashed it.

Everything felt all right. She didn't want to open the backpack in the rain to find out. After selecting a sturdy branch above her, she used the straps to secure the backpack. She took a flashlight from one of the outside pockets.

For the moment, no sign remained of the cliff as the water continued to surge through the jungle and inland.

That's got to be at least fifteen feet deep, Annja thought. She tried to remember how high the lowest tree branches had been from the ground. Then she realized she didn't know where the lowest branches were anymore.

"How long is this going to last?" someone above her asked.

Annja looked and saw Jason Kim sitting a few branches up. He clung to the tree bole. A young German woman had her arms wrapped around his waist. Both of them looked terrified.

"I don't know," Annja said. "Could be only a few minutes. Might possibly be hours."

"Is it over?" the young woman asked.

Annja was hesitant to answer. "I think so." Given the amount of water that had flooded the land, she knew that whatever had happened at sea to cause it had to have been powerful.

The massive tree swayed under the constant bombardment of the waves. They were lessening, but still dangerous.

Lightning burned through the night and revealed the dark clouds swirling and twisting overhead. The harsh peal of thunder came immediately.

Another crack, this one different from the thunder, issued from the left. A chorus of yells and cries for help followed.

Turning in the direction of the voices, Annja spotted a teak tree as it fell into the water. Three dig members clung to the branches as it went down. Annja guessed that the tree had been poorly rooted or had rotted and weakened. Either way, the surging sea started to carry the tree away.

"They're going to get killed," Jason Kim said.

Other people voiced the same concerns.

Annja knew that death was a possibility. If they stayed with the tree, if they didn't get smashed by the branches, they might survive. But the water might carry them a mile or more into the interior, just far enough for them to get lost and possibly die from some other cause.

She grabbed her rope and shimmied along the thick branch she was on. Just as the branch started to sag beneath her weight, she jumped forward for the next tree. The teaks overcrowded the area and the branches grew close together.

By the time she caught a thick branch in front of her, she'd already chosen her next landing point. Like an aerial gymnast working uneven bars, she made her way through the trees faster than the floodwaters could carry away the huge tree. She also got closer to the water level. Her hands burned from friction against the bark.

After her fifth jump, when she knew she was out of trees, Annja shrugged the rope from her shoulder. Setting herself on a limb as wide as her body, she shook out the rope, swung the grappling hook and cast.

The grappling hook landed in the branches of the fallen tree. It jerked and bounced as it slid along the length of the tree without securing a hold.

C'mon, Annja thought. Take hold somewhere.

The grappling hook snugged up against a thick branch. Annja yanked on it like fishing line to set it. Satisfied it was securely in place and knowing that she'd never be able to hold the tree on her own, she dropped over the back of the branch she was on and paid out rope as she plunged into the water.

For a moment as she entered the water, she was afraid. The drop was little more than six feet, but she knew anything could be under the water. If she was knocked unconscious or seriously injured, no one would be able to help her.

The flashlight beams of the other dig site members played over the water as they tracked the tree caught in the surge. The glowing light continued moving away from Annja.

When the rope bit into her hand, Annja paid out more line and fought the current to get to the base of the tree she'd dropped from. Once she had hold of the tree, she pulled herself around it and looped the rope. Then it, too, became a deadly threat.

If she got caught in the rope, if the weight didn't amputate her fingers or a hand or break them, it might trap her below the water and leave her to drown. The coarse fiber burned along her palm.

The rope pulled taut. The tree she'd attached it to shivered under the assault. But it held.

Working quickly, Annja tied the rope off and made it secure. She had to work one-handed while she held on to the tree with the other. Then, as black spots danced in her vision from lack of oxygen, she kicked and swam up next to the tree.

The flashlight beams from the other dig site members barely reached the fallen tree. In the dim light, Annja saw that all three people still held on to the branches less than thirty feet away from her.

Annja abandoned her hold and let the current take her. The current wasn't as strong as it had been earlier.

Swimming in it was difficult but it was only a short distance.

When she reached the tree, she hung on for a moment to gather her strength. Instead, the constant battle with the current only leeched energy from her. She forced her body out of the water and onto the tree.

"Is everyone all right?" Annja shouted above the noise of the storm and the rushing water.

All three college students, two female and one male, nodded. All of them looked pale and frightened in the flashlight beams and the lightning.

Tethered at the end of the rope, the tree danced and jerked like a fish on a line. Annja spotted the white scars left in the bark by the grappling hook's prongs. She could see the branch the hook had caught had started to tear away.

Annja made her way across the slippery tree trunk and grabbed the branches from another nearby tree. She held tight and saw blood from the cuts on her hands. She ignored the pain and kept gripping.

"Get into the tree," Annja ordered the others.

At first none of the three college students wanted to move. All of them were from Lochata's university, and they all spoke English.

"Now!" Annja commanded in a more forceful tone. "That branch is going to tear free. I don't have another rope and I don't think we'll get this lucky twice."

One of the women spoke to the others in her native language. She got them up and moving. Awkwardly and fearfully, they made their way into the other tree.

Annja helped them, then pulled herself into the

branches. She felt the cold from the storm splintering through her body.

Postadrenal surge, she told herself as she hunkered down and rubbed her arms. You'll sleep well tonight. If you find a place to sleep.

The storm continued unabated. A few minutes later, the broken tree tore free from the rope and floated away. It collided with several other trees before disappearing into the darkness.

Annja settled in and got as comfortable as she could. It promised to be a long night.

ANNJA WOKE with the dawn. The sun painted the eastern horizon pink and purple with hints of gold and ruby. Blinking against the brightness, Annja relished the increasing warmth. When she pushed herself up from the crook of the tree's branches where she'd slept, the pain in her hands reminded her of the damage she'd done.

She looked down at them and found several tears and scrapes across her palms. They weren't as bad as they'd felt last night, but they were still painful when she flexed them.

"Professor Creed, I can't believe you slept like that."

Annja glanced up and shaded her eyes against the sun. One of the Indian college students sat on a limb above her. She was young and thin with long black hair. Annja tried to remember her name and finally got it.

"Indira, right?" Annja asked.

The young woman nodded. "I couldn't sleep a wink."

"I probably shouldn't have." Annja looked down. The water level had dropped considerably, but it still looked several feet deep. There was no sign of the campsite or the vehicles.

"I left my computer down there," Indira said. "All my stuff." She bit her lower lip. "It's probably ruined, isn't it?"

Annja hated giving the young woman the bad news. "I'm afraid so."

Tears filled Indira's eyes.

A guilty feeling stole through Annja even though she'd had nothing to do with the tsunami. She looked back at the tree she'd originally climbed. Her backpack still hung safely from the limb. She sighed in relief. Replacing the equipment would have been a pain, but financially she could have done it. However, getting replacements could have been difficult.

She grabbed the limb over her head and pulled up. Her hand burned as the cuts pulled. A quick inspection revealed that none of them had broken open. Infection could be a problem, she told herself. And the first things you take care of when you're out in the tropics are your hands and feet.

She discovered she was sore from sleeping in the tree. All the diving, jumping and swimming had probably contributed to it, she thought.

"Professor Creed?"

Still not used to the formality, Annja thought. She wasn't actually a professor but Lochata Rai had told all of her charges they were to address Annja that way.

The speaker was the male college student. Annja felt bad that she couldn't remember his name at all.

"Yes?"

"I just wanted to thank you for saving us," he said. "That was the coolest thing I think I've ever seen done."

"WE WERE VERY LUCKY," Lochata said. "Everyone survived the experience."

"I know. But the flood destroyed the dig site." Annja waded through the hip-deep water and felt the pull of the flood's withdrawal. The sea continued to return to its proper boundaries.

Annja and the professor had organized the dig members into teams responsible for searching for supplies that might have survived the flooding. Prizes turned up with hopeful regularity, though many of them were farther inland. A lot of the food and water was in waterproof containers. Unfortunately, many of those containers were buoyant. The deluge had ripped the tents free of their stakes and allowed them to be carried inland or back out to sea.

"What was buried under the earth will still be there when we are ready to begin again," Lochata said. She reached forward and plucked a snake from the water, examined it for a moment, then hung it on a tree limb.

"Think it'll find its way home again?" Annja asked.

"Or make a new home, perhaps." Lochata watched the snake slither along the branch until it found a place in the sun. It coiled up and sat there.

One of the male students sang out joyously fifteen yards away. He hoisted a bottled sports drink into the air as if he'd just won an Olympic event. He spoke in his own language.

"He says he's found a whole box of the sports drink," Lochata translated with a smile. "At least forty-eight bottles."

The find drew the others to the area and they fanned out to search the underbrush for more food and drink.

Annja knew they weren't going to starve. Her satellite phone had allowed Lochata to get in touch with rescue centers in Kanyakumari and request assistance.

According to the dispatch officer Lochata had conversed with, the city hadn't been hit by the tsunami. The disaster seemed to be fairly localized, but several small villages had been hard-hit, as well.

"Are you going to continue the dig?" Annja asked.

"If I'm able. I still have to contact the university."

"It seems a shame to walk away from it now. We've just gotten started," Annja said.

"I agree."

"And it's not likely there'll be another tsunami."

"I don't think so, either."

Annja watched the university students splash around in the water. "Do you think many of your interns will stay on?"

"I can only ask." Lochata raised her thin shoulders and dropped them. "For many of them, this will be a grand adventure by tomorrow. Something they'll be able to brag about to their classmates when we return to university. However, they have to return in two weeks no matter what. That's all I could arrange for them to be away from their studies. I was not able to schedule this for the summer due to the monsoon season."

"You can always threaten them with their grades." Annja smiled.

"Hey!" someone shouted. "I found gold!"

The unusual cry drew Annja's attention at once.

Across the water, brown and thick with dirt and debris, one of the male college students held up an object that held the dark yellow luster of gold. He had to use both hands to hold the object.

Lochata and Annja trudged through the water and joined him.

"Let me see that." Lochata took glasses from her vest pocket, slipped them on, then reached for the object the young man held.

Annja peered over the diminutive professor's shoulder for a better look.

The object was hardly larger than Annja's closed fist, but it was too heavy to be common metal. It looked like an egg, elliptical in shape. But at the top a fist poked through.

"What is it?" someone asked.

"It appears to be a mechanism of some sort," Lochata answered.

"Is it gold?" someone else asked.

"I believe so, yes." Lochata's fingers glided around the figure.

"Where did you find it?" Annja asked the student who'd found it.

He pointed at the calf-deep water. "Here. I stubbed my toe against it. I figured it was just a rock, but when I looked down I saw that gold color. When I picked it up, that's what I found."

Several of the students took renewed interest in the surrounding area.

"Am I going to get to keep it?" the student asked.

"Dude," Jason said, "if I can't keep one lousy skull

out of the dozens we found, there's no way they're going to let you keep a solid-gold paperweight."

"It's not a paperweight," Annja said.

"Then what is it?"

"No one makes a paperweight out of solid gold," one of the female students said. "Except maybe Paris Hilton or Britney Spears."

Annja ignored the chatter. She watched as Lochata's fingers found the hidden release. The mechanism inside the egg-shape whirred to life. The device split open like the sections of an orange to reveal the figurine inside.

It was a woman.

At least, part of it was a woman. From the waist up, the fantastic creature was a woman. She held one fist above her head. In the other she held a short whip.

But below the waist she was a snake. Her serpentine half sat in a tight coil and balanced her.

4

"A ship! I see a ship!"

Wearily, Goraksh Shivaji lifted his head and stared out at the bleak expanse of the Indian Ocean from the deck of his father's ship, the *Black Swan*. He'd barely managed a handful of catnaps during the night.

He should have been home in Kanyakumari studying algorithmic design paradigms. The professor this semester was harsh. College life wasn't easy for him. It didn't help that his father expected him to work a fifty-hour week in the warehouse.

Over the past two years of his career at university, Goraksh had thought about telling his father that he was quitting the warehouse. But he needed the pittance his father paid him to pay his tuition.

Jobs were hard to come by, especially ones that worked around a college schedule. Also, working in the warehouse guaranteed that he could live in his father's house. If he was on his own, he knew he wouldn't be able to make ends meet.

As it was, when Goraksh finally graduated, he was going to owe a small fortune to the university. He would have his degree in computer science. Then he would be able to get a good job in the United States, maybe designing video games, and finally leave his father's warehouse behind for good.

But that was the dream. Tonight was all about working for his father. If you could call piracy work, Goraksh grumbled sourly.

"Goraksh, do you see the ship?" His father's voice was stern. Rajiv Shivaji was a hard, lean man in his early fifties. He wore the turban and steel bracelet—the *kara*—of the Sikh, and his beard was full. He also carried a .357 Magnum revolver in a shoulder holster.

"Not yet, Father," Goraksh replied. He held the high-powered binoculars to his eyes and swept the surface of the sea. The light hurt his eyes. He swayed to the rise and fall of the waves as the cargo ship strained under full sail.

Rajiv stood at the prow of the ship and held on to the railing. Goraksh had never seen a man more able who had taken to sea. It was easy to imagine him sailing with the likes of Sinbad the Sailor and other heroes.

Except that Rajiv wasn't a hero. He was a pirate and a thief, and he had set sail with his crew after learning the tsunami had struck. They'd expected to find several ships swamped at sea. So far they'd found none.

"Fyzee," Rajiv yelled up to the old man standing in the wire crow's nest twenty feet above the pitching deck.

"Yes, Captain."

"Do you still see the ship?"

"I do. It's only a short distance away." Fyzee pointed. He was old and potbellied. His beard and hair had turned snow-white long ago.

Goraksh followed the direction the old man was pointing, then lifted the binoculars to his eyes again. This time he saw the ship. He knew then why he'd lost it—the ship was upside down.

Judging from the rough, unadorned exterior and the barnacle-covered hull, the craft was a cargo ship. It was one of the lunkers that local businesses used to cross the Indian Ocean on regular routes. They were operated for a song and only required a skeleton crew. Goraksh thought of the ship's crew and wondered what had happened to them.

Sickness lurched through Goraksh's stomach when he thought of how cruel the sea could be to those who were lost in it. He'd been with his father when they'd reclaimed bodies from the ocean. Sometimes, after the sharks had gotten at them, there were only parts of bodies. But they'd inspected them for anything worth stealing and quickly shoved the gruesome remains back into the sea.

"Well?" his father demanded.

"I see it," Goraksh replied.

"Where is it?"

"South by southwest."

Rajiv called orders back to the helmsmen. The crew came about sharply as the ship took on a new heading.

"Are there any survivors?" Rajiv asked.

"None that I can see." Goraksh kept scanning the

boat from prow to stern. He knew they weren't looking for survivors. Anyone who had lived through the storm would only complicate things.

Rajiv gave orders to trim the sails. Goraksh put his binoculars back in their protective case. Tension knotted in his stomach when he thought of what might lie in the overturned ship's hold.

GORAKSH BRACED HIMSELF as the ship came alongside the cargo vessel expertly. Tires tied along the length of their port side muffled the impact.

"All right," his father growled as he paced the ship's deck, "get aboard and discover what the gods have favored us with on this trip." He stopped in front of Goraksh. "Go with them, college boy. See how a man dirties his hands to put food on the table."

Goraksh wanted to argue but he couldn't meet his father's gaze. His father had been angry with him ever since Professor Harbhajan stopped by the warehouse early in the week.

The warehouse had been full of stolen and illegally salvaged items. Fortunately the professor hadn't recognized any of it. But Goraksh's father hadn't let him forget that the professor could just as easily have turned them in to the police.

Professor Harbhajan had graded the projects his class had turned in at the start of the semester. He'd stated that he'd been particularly impressed by Goraksh's work. His father had been incensed when he'd heard about the visit and the topic.

Rajiv was one to hold grudges for years. Goraksh

knew that no matter how long he lived he would never be forgiven the trespass he himself had not caused.

Without a word, Goraksh nodded. He kicked off his shoes and clambered over the ship's side with the rest of the boarding crew.

UNDER THE HOT SUN, Goraksh held the battery-operated saw and worked quickly. He'd paired up with Karam, one of his father's oldest crew. The man was emaciated by age and alcoholism. His gray beard showed stark against his dark skin. Old scars inscribed leathery worms against his features.

The saw jumped and jerked in Goraksh's hands as he held it to the task. The blade chewed through the wooden hull and threw out a constant spray of splinters. He remained aware throughout of the ship's erratic movement in the water.

Finally, when he had a square cut that measured a yard to a side, Goraksh pulled the saw back and stomped his foot on the square. The section dropped down into the hold. Goraksh heard it hit water only a short distance down.

"There's water in the hold," Karam called across to the other ship.

Rajiv leaned on the railing. "Find out what else is down there."

Karam nodded.

"Goraksh," Rajiv called. "You'll go inside."

For a moment Goraksh thought of disobeying his father's order. His father knew he had a fear of enclosed, dark places.

"She's the *Bombay Goose*," Rajiv said. "I checked her manifest."

Goraksh knew his father paid someone off in the customs house for ships' manifests.

"She's carrying electronics," his father continued. "Computers, DVD players. Those will sell nicely on the black market."

They're probably all destroyed, Goraksh thought. But he knew better than to point that out to his father. Rajiv Shivaji always believed good things would happen to him.

Rajiv looked over his shoulder and shouted for the scuba gear to be brought up from the hold.

Karam caught Goraksh's eye and spoke in a low voice. "Go slowly, boy. Everything will be all right if you just go slowly."

Goraksh nodded but he didn't believe it. He didn't think for a moment that the crew had gotten off the ship in time. He only hoped that they'd all been lost to the sea.

WITH THE AQUALUNG STRAPPED to his back and an underwater floodlight in one hand, Goraksh dropped into the ship's hold through the hole he'd cut. He was in total blackness except for what little light entered the hold through the cut-away hole.

He stayed submerged for a moment and blew into his face mask to equalize the pressure. Then he shone the floodlight around the hold. Boxes lay on what had been the hold's ceiling or floated in the water. The air pocket between the hull and the waterline was less than three feet deep.

Goraksh didn't know what was keeping the cargo

ship afloat. Thinking like that made him nervous, though. If the ship suddenly went down, the sea bottom was nearly a half mile down. If he didn't get out quickly enough, it would take him with it.

Don't think about that, he instructed himself. Get the job done.

He surfaced and shone the floodlight up at Karam. "Send down the net."

Karam nodded and dropped the cargo net down. Other members of the crew used battery-operated saws to widen the hole in the hull.

Goraksh grabbed a fistful of the rough hemp strands and pulled the net under with him. He selected a crate at random and wrapped the net over it. Then he yanked on the rope to signal Karam and the others to haul it out of the hold.

An arm settled around Goraksh's neck and shoulder. Fear ripped through him as he flailed in the water with his free hand to turn around. He aimed the floodlight behind him and instinctively centered it on the figure.

The dead man's mouth and eyes were open. Yellowed eyes and yellow, crooked teeth showed.

That was all Goraksh noticed before he screamed in terror and tried to swim backward. The respirator dropped from his lips and his face slammed into a sus-pended crate hard enough to almost knock him out. He swallowed seawater as he tried to breathe, then remem-bered he was underwater.

Fighting the panic that filled him, unable to get the dead man's face out of his mind, Goraksh dropped the floodlight and used both hands to shove crates away from him so he could reach the surface. He pushed off on a floating crate and got enough lift to reach the edge of the hole that had been cut in the hold.

Sick, barely able to breathe because of his fear of dead things and the seawater he'd swallowed, Goraksh hauled himself out of the hold. He couldn't stand and ended up on all fours as he retched out the seawater.

When his stomach finally settled, Goraksh felt drained and embarrassed. He forced himself to his feet and stood on shaky legs amid the mess he'd made.

"Are you through shaming me?" his father roared from the other ship.

Goraksh faced his father and intended to speak roughly, as a man would do. But his words were soft and without direction.

"The crew went down with the ship," he said.

"Good. Then maybe they didn't have time to call in this location," his father said. "Maybe we'll have more time to work."

Even after all the years he'd lived with the man, Goraksh couldn't believe how callous he was. Rajiv had brought Goraksh along on the pirating expeditions after storms for eight of his twenty years. During the past four, Goraksh had been expected to take part in stealing whatever cargo they could salvage.

Finding the illegal salvage was one thing, but getting away with it was quite another. The Indian navy and merchant marine, the British navy and the International Maritime Bureau, were all problems. Rajiv Shivaji considered those risks a part of doing business.

Goraksh recognized them as an end to the life he

wanted. His father was a pirate. Rajiv Shivaji carried on an old family enterprise. Goraksh never romanticized the nature of what his father did.

But if Goraksh was ever caught doing his father's business, he knew his dream future was forfeit. Still, he loved his father. After his mother had died, his father had raised him and had never taken another wife. It had only been the two of them.

If Goraksh was ever to be asked if he feared or loved his father more, though, Goraksh didn't know what his answer would be.

KARAM USED a crowbar to open the crate Goraksh had selected from those in the flooded hold. Water, foam peanuts and boxes of iPods spilled out across the ship's hull.

"They're ruined," Rajiv snarled. "Go below and find something salvageable."

Goraksh put the respirator back in his mouth and dived back into the hold. He recovered his floodlight and tried not to look at the dead man floating amid the boxes. Then he found two more.

He bagged more crates and sent them up. During the time he waited for the net to be sent back down, he scouted the hold. Two hatches, one at either end, normally allowed access to the upper decks. Both of them had jammed.

If there was anything in the crew's quarters, they wouldn't be able to get to it without cutting through the floor or forcing the hatches. Goraksh hoped his father wouldn't demand that. Doing either of those things might upset the equilibrium of the ship.

Even now he truly believed the ship had sunk lower in the water. He reached the opening they'd created more easily.

Pounding echoed throughout the hold. Goraksh felt as though he were trapped in a gigantic drum. He netted a final crate, thinking his efforts were going to be as wasted as the other times. He surfaced.

Karam leaned down into the hold. He cupped one hand around his mouth to be heard over the sound of the sea against the hull. "Your father wants to leave."

"All right," Goraksh responded. He swam through the maze of boxes to the opening and wondered what had made his father change his mind. Not even the fact that they'd only pulled up ruined electronics in over a dozen attempts would have made Rajiv Shivaji give up on the hope of turning a profit.

Something had happened.

"IS ANYONE OUT THERE? Can anyone help us?"

Goraksh stood beside his father in the ship's wheelhouse and listened to the broadcast over the shortwave radio. His sodden clothing gave him a chill.

"Hello? Hello? God, please let someone be out there. We need help. Our boat is sinking. Please. *Please!*"

The voice belonged to a woman. She sounded young and frightened.

Rajiv glanced at the radio operator. The man worked quickly with a slide rule, compass and map. He made a few tentative marks and watched his instruments again.

"Why don't you answer her?" Goraksh asked. For a

moment he couldn't help imagining his girlfriend at the other end of the radio connection. Then again, Tejashree feared the open ocean and wouldn't accompany him sailing.

"Because I don't wish to answer her," Rajiv said.

Goraksh fell silent and knew better than to ask again.

"Our boat is the *Grimjoy*," the woman said.

Although he tried, Goraksh couldn't decide if her accent was American or Canadian. He knew there was a difference between the two, but he didn't quite know how to tell. He would have known if she had a British inflection.

"*Grimjoy*," one of his father's men said as if he were familiar with the vessel.

"I know." Rajiv nodded happily. "I know that boat." He looked at the radio operator. "Can you locate it?"

The man made a few final notations on the map. "I have it now." He handed up a slip of paper with the co-ordinates listed.

"How far away are we?" Rajiv demanded.

"Ten or fifteen miles. They're north of our position."

"Is the boat in the open sea?"

The radio operator shook his head.

Goraksh knew that within the country's boundaries the authorities would arrest his father for what he was doing. Most of the men on the *Black Swan* had been in trouble with the law on some occasion.

"Does anyone else know they're out there?" Rajiv asked.

"I've been monitoring this frequency. So far they've received no reply."

"Good." Rajiv gave the paper with the coordinates to the helmsman. "Set a course to take us there immediately."

The man nodded and hurried away.

Rajiv strode out of the wheelhouse and onto the deck. He bellowed orders to abandon the sinking cargo ship and put on sails.

Goraksh watched his father, but he listened to the woman's plaintive voice coming over the radio frequency.

"Please. Someone has to be out there. We're adrift. I don't know how to work the boat."

In seconds the *Black Swan* got under way. She heeled hard to port, caught the wind and sliced through the rolling waves like a thoroughbred.

When he joined his father on the deck and saw the savage exuberance on his father's face, the sick knot inside Goraksh's stomach twisted more violently. He'd never seen his father kill anyone, but he was aware of the stories that were told in the rough bars and opium dens in the darkest corners of Kanyakumari that said Rajiv Shivaji was a murderer several times over.

5

"Dude, *nagas* were evil."

"Maybe in Dungeons & Dragons Third Edition, but not in Three-Point-Five. In Three-Point-Five you could roll up a *naga* character and play one. You could be Lawful Good if you wanted to."

"Yeah, well Three-Point-Five ripped D&D's canon all to hell. It was just a stupid marketing ploy to bring back players who wanted to play monster characters and got pissed because they couldn't."

"Playing monster characters is cool."

The constant chatter had finally gotten on Annja's last nerve as she scanned the ocean shallows for more artifacts like the *naga*. She'd been listening to the arguments cycle viciously between Jason and one of Professor Rai's students for two hours. At first the discussions had been amusing. Now they were exhausting.

"Hey!" Annja turned around quickly and brought

both of the younger men up short. They splashed to awkward stops in the water. "Gamer geeks—enough with the chatter."

Jason and the other young man just looked at her owlishly. They even blinked at about the same time.

"The *naga* statue we found isn't a playing piece from some long-lost D&D game," Annja said. "We're supposed to be out here looking for more artifacts."

She and Lochata had agreed to keep the students busy until the rescue helicopter arrived. They'd salvaged enough water and energy drinks to get them through the next few hours.

"You think that *naga* was like part of a chess set?" the other young man asked.

Irritated, Annja pinned him with her gaze. "What's your name?"

"Me?" The young man pointed at himself and looked surprised.

"Yes. You."

He shrugged. "My name is Sansar."

"Fine. Listen up. No, I do not think that *naga* statue was part of a chess set. Or any kind of game."

"It would be kind of big, I suppose."

"Sansar," Annja said, struggling to maintain her composure.

The young man looked at her.

"What I think is that the *naga* statue came from somewhere out there." Annja waved at the shallows that lapped at the foot of the cliff. The flooding had almost totally receded.

"Like, it was just laying out here somewhere?"

"Or was buried under the sand." Sand, shells and other debris from the sea lay strewed across the dig site and into the jungle. "The tsunami moved a lot of sea floor. Maybe it shifted some things around," Annja said.

"You think there's more out here?"

"I think there *could* be more out here," Annja corrected. She couldn't believe how lackadaisical the two were about potentially finding more artifacts.

"So we could be out here tromping around in the water for no reason," Jason said.

"Personally, I think it beats sitting around the dig site in muddy clothes waiting for help to arrive," Annja said.

Jason frowned. "If my PSP hadn't gotten washed away, I'd rather be sitting in the shade playing a game."

Okay, Annja thought with a sigh, forensic anthropology in a nice, quiet lab is *soooo* going to be your thing.

"When's the rescue helicopter going to be here?" Sansar asked.

"I don't know," Annja answered. She felt a headache coming on, but she didn't know if it was caused by hunger, the hot sun, spending the night in a tree or listening to the never-ending argument.

"Man, I hope somebody finds more food," Sansar said. "Do you think a Pringles can could survive getting submerged? I mean, if it hasn't been opened. Those things are watertight before you peel them open."

Annja turned to face the two. "I've got an idea."

They waited.

"Why don't you two walk in that direction?" Annja pointed in the opposite direction.

Jason looked that way, then he looked back at Annja.

"Why do we have to walk that way? Why can't we walk with you?"

"Because we can cover more ground if we separate." Annja hoped she sounded reasonable instead of frustrated and resentful of the company she was keeping.

"Yeah," Jason said, "I can see that. But why do you have to have this end? Why can't we have it?"

Annja stared at Jason. "We've been through a tsunami. We're trapped out here without supplies. And you want to argue over which end of the Indian Ocean we're going to search for artifacts that could be harder to find than a needle in a haystack?"

Jason's self-preservation suddenly kicked in. He held his hands up before him. "Hey, you know what? This end is just fine with me." He looked over his shoulder and faked smiling happily.

"What if she's telling us to go that way because she believes she's going to find something *this* way?" Sansar said suspiciously. "How do you know she doesn't just want all the glory for herself?"

"Dude," Jason whispered, "you should really keep your mouth shut about now. She could kick your butt."

"Okay," Annja said as evenly as she could, which she knew wasn't very even at all, "you guys take this end. I'll take that one." She pulled the straps on her backpack and headed the other way.

"You just really made her mad," Jason told his companion.

"Me? You're the one that started the argument over the D&D rules."

Annja tried to block them from her hearing, but she

was doomed to failure because sound carried more clearly and farther over water than it did over land. There were times when she preferred working alone on a dig. This was one of them.

Doing a field study with Professor Rai was a treat. The woman had traveled extensively around the Indian subcontinent and been part of every major dig the Archaeological Survey of India had done in the past twenty years. Annja knew she could learn a lot. She also knew that the professor had played up Annja's involvement to the local papers to get more press due to the *Chasing History's Monsters* connection.

The Shakti-sacrificial-victims dig hadn't been set up to ferret out any new information. It was fieldwork designed to season the professor's class and to provide more substantiation to the book Lochata was writing on Shakti.

The gold *naga* statue was a totally unexpected find. Annja just hoped there would be more. She didn't see how there couldn't be.

Jason and Sansar kept up their argument, though at a lower volume. They obviously weren't paying attention to what might be in the shallows.

Annja sighed unhappily. She was wet, hungry, tired and pushed to the breaking point of her patience. She wondered how the two students could be so completely useless. She wanted to find another artifact to show them what could happen if they actually applied themselves to the task at hand.

"Hey!" Jason yelled with sudden enthusiasm. "Look what I found!"

"So how old is it?" Jason wanted to know.

Standing in the shallows where the artifact had been found, Annja upended the fired clay pot and studied the bottom. "49 B.C.," she said

"Wow," Sansar said. "That thing's over 2050 years old."

Jason slapped him on the back of the head. "She's goofing you. How would a potter know that he made a pot forty-nine years before Christ was born?"

"Oh." Sansar rubbed his head. "I knew that. I was just so excited over finding it that I wasn't thinking."

"You didn't find it. *I* did," Jason said.

"We were walking together. That means we both found it."

"I seem to recall bending down to pick it up from the water," Jason replied.

"Both of you can shut up," Annja suggested.

They looked at her, clearly offended but silent nonetheless.

"I think you have found something," Annja said a short time later. "It even ties in with the dig Professor Rai has initiated." She pointed to the figure of a six-armed woman riding a tiger.

"Shakti, right?" Jason asked.

"Right."

"I thought I recognized her."

"This lays out some of her story." Annja slowly turned the pot to display the collection of images around the base.

The images were sculpted to lead one into the other. The image next to the one of Shakti on the tiger showed her at court with several ladies-in-waiting fanning her. Still another showed her in battle with Shiva, her lover.

The final image showed her sacrificing herself on a funeral pyre to Shiva.

"Makes you wonder how long the Shakti cult was here," Jason said.

"It does," Annja admitted. "But it also makes you wonder how wide the belief in her was spread."

"What do you mean?"

"Did this come from the dig site? Or was it brought in from the sea?"

"You think someone threw it away?"

"No." Annja struggled for patience. "I think the *naga* and the pot could have been part of a ship's cargo."

"Cool," Sansar said. "You mean you think there's a sunken treasure ship loaded with gold out there?"

"No," Annja said. "I don't."

BUT THE OTHER MEMBERS of the dig site were quickly convinced by Sansar and Jason that the ocean shallows were burgeoning with gold just waiting to be scooped up. They'd taken a break to go get bottles of water and quickly spread the news of their find. When they'd returned, most of the dig site members had returned with them.

The students split into groups and prowled the water like children on an Easter-egg hunt. Jason and Sansar had stopped arguing long enough to locate a fishing net that had washed up. They weighted the bottom with stones and were dragging the ocean bed.

Annja reluctantly admitted to herself that the two were definitely inventive.

"Not exactly the most organized effort, is it?" Lochata asked.

"Not even," Annja agreed. Her headache had gotten worse. Despite the pain and the frustration she felt, she worked in the journal she was keeping for the Shakti dig.

She sketched the bay area's general geographical characteristics and marked the site where the clay pot had been found. The spot where the *naga* statue had been found had already been marked.

"I'm surprised the pot survived the tsunami," Lochata said.

"Not to mention hundreds or thousands of years at the bottom of the ocean," Annja said.

"It wasn't there thousands." Lochata turned the pot carefully in her hands. "This was kiln-fired."

"So it came from a city or a town," Annja said.

Lochata nodded.

Annja flipped back through the notes she'd made prior to boarding the plane in New York. "The closest city I know of that was on the coast within that time frame was Kaveripattinam."

"There were a few others. Smaller, but still viable. But it was Kaveripattinam that the world came to see and trade with. Until a tsunami destroyed much of it twenty-five hundred years ago," Lochata said.

"We're a long way from Poompuhar," Annja pointed out. Kaveripattinam had been rebuilt over time, though so much of the ancient architecture had been lost, and it had been renamed Poompuhar.

"The pot could have come from a merchant ship, then," Lochata said. "I've worked with a lot of the pottery that was found offshore there. This piece looks like other pieces that were recovered there."

"Even the bas-relief?"

"No. I was talking about the composition of the materials and the technique used to fire it." Lochata ran her fingers over the raised images of Shakti. "These mark the pot as something other than an everyday pot. This was probably intended for a religious service. Or as a cherished gift for a lover or a family member."

Annja showed the professor her drawing. "The pot and the statue were found in a relatively straight line."

Lochata nodded. "I'd noticed that."

"It would probably help if some of the students searched deeper into the jungle. Anything that was light would have washed farther up the shore."

"When I can get them to stop looking for gold," Lochata said, "I intend to have them search there." She sighed. "Provided they're interested in continuing the dig."

Annja glanced out at the students walking through the shallows and smiled. "I think they're interested. We just need to find a few more things to keep them that way."

WHEN ANNJA STRIPPED DOWN to her bikini she claimed the instant attention of every male in the dig crew. She felt a little self-conscious as she walked toward the water.

She had a good body. She knew that. Hours of work on the weight machines and StairMaster, hours spent in the boxing gym she frequented and an active lifestyle guaranteed that.

And the bikini showed off her figure. She'd worn it

under her clothes so she could go for a quiet, private swim in the ocean at the end of a long hot day in the pit.

The snorkel and swim fins she carried were borrowed from one of the students whose belongings had turned up in a tree. At the water's edge, she sat on a rock, pulled the swim fins on and settled the mask over her face. She tried to ignore the continued staring as she made her way out into the water.

She swam out twenty yards or so. From the way the seabed gradually sloped out, she guessed she was in fifteen to twenty feet of water. After a final deep breath to charge her lungs, she dived.

The crash of the surf against the cliff suddenly seemed distant. Annja felt as if she'd been wrapped in cotton. She swam cleanly as she moved her arms and legs almost effortlessly.

The ocean was clearer than she'd expected. With the disturbance caused by the tsunami she'd anticipated a lot of debris in the water. There was a lingering fog, however, that limited her visibility. She resisted the impulse to clean her face mask.

As always, the beauty of the sea overcame her. The brilliant colors of the fish in the tropical saltwater environment caught her eye again and again. Schools swam and darted in unison. Several coral growths stood proudly on the sea bottom. An eel whipsawed through less than a dozen feet away.

You're not here on a sight-seeing tour, Annja reminded herself. She swam down to within reaching distance of the seabed.

She hadn't swum far when she found the first gold

coin. She dug it out of the loose sand and spotted three more.

In the excitement, she hadn't paid particular attention to the tightness that strained her lungs.

When she flipped over to begin her ascent, she noticed the hull of a speedboat cutting through the water toward the shallows. She surfaced and spit out the snorkel mouthpiece, breathing deeply to replenish her depleted lungs.

The boat moved in too close and too quick. Several students had to flee the water. Four men sat in the speedboat. They laughed at the students and mimed the panicked reactions of some of them.

Annja treaded water on the other side of the speedboat. She scanned the craft and noticed the name and registration were missing or covered over.

Things didn't look good.

One of the men brought up a bolt-action rifle and shouted something in his native tongue. Another man tapped him on the shoulder and spoke quickly.

The man with the rifle addressed the dig members again in English. "I want to talk to your boss now or I will start shooting."

6

The *Grimjoy* rocked on the sea with a careless abandon that told Goraksh the craft hadn't been properly anchored.

The yacht was a thing of beauty. At least forty feet long, the boat was a shipbuilder's confection of polished teak and brass. It was also rigged and powered to be a motorsailer, capable of traveling with the wind or by the big engines.

Goraksh listened to his father's bellowed commands and helped with the sails as the *Black Swan* closed on the yacht. The lookout in the crow's nest relayed that no one else appeared to be about.

Grabbing his binoculars, Goraksh studied the yacht. He spotted a red-haired woman in a bikini waving frantically in the stern, but no one else appeared on deck.

"What do you think?"

His father's unannounced presence at his side startled

Goraksh. He took an involuntary step away before realizing it was his father.

"What do I think about what, Father?" Goraksh asked.

Rajiv nodded at the yacht. "It could be a trap."

"A trap?"

"There could be armed men belowdecks waiting till
we're within range," Rajiv said as calmly as though
they were discussing the prevailing winds. "They could
have rifles or machine guns. Perhaps even a rocket
launcher. Those things are not as hard to get hold of as
they once were."

Goraksh knew that; his father sometimes dealt in mu-
nitions. But everyone who had a boat and needed money
did. There were always rebel forces in India, Africa and
the Middle East who needed them. Sometimes Rajiv
only hired out to transport someone else's weapons.

The woman continued waving and yelling.

"I don't think it's a trap," Goraksh replied. "The
woman appears too afraid."

"Perhaps you're right," his father said. "It pays to be
right." He paused. "But it also pays to be careful." He
barked an order to one of the men.

Instantly the order was relayed to the other men. All
of them armed themselves with assault rifles that were
brought up from belowdecks. Possession of any one of
the weapons was enough to get them in serious trouble.
Having all of them—

Goraksh swallowed hard. He didn't know what
having all of them meant. But it couldn't be good.

The woman didn't think so, either. She shrank back,
then turned and fled into the cabin.

"Here."

Goraksh turned once more to his father. Rajiv held a semiautomatic pistol in his hand.

"Take this in case you need it," his father commanded.

Reluctantly, but trying hard not to show it, Goraksh took the pistol. The weapon fit his hand instinctively, but it was a lot heavier than he'd expected. He prayed he wouldn't need it.

Rajiv gave orders to close in.

THE *BLACK SWAN*'S CREW lashed their ship to *Grimjoy*. Then, after they pulled on disposable gloves to prevent leaving fingerprints, they followed their captain aboard.

Goraksh accompanied his father because Rajiv grabbed his shirt and propelled him forward. The pistol dangled at the end of Goraksh's arm. He wasn't even sure if the safety had been switched off.

The *Grimjoy*'s deck rocked beneath their feet. Waves slapped flatly against the ship's hull.

"Do you know why I brought you last night?" Rajiv whispered into Goraksh's ear.

"No."

"Because you are twenty," his father whispered. "Because you are a man. And because the men who work for me wonder why my son—my only son—hasn't taken his place with me."

Goraksh went forward to the ship's cabin afraid he was going to be shot at any moment. He thought he might be sick.

"You are a Sikh," his father whispered vehemently.

"The blood of warriors runs through your veins. *I* put it there."

Goraksh stood at his father's side in front of the cabin door. He heard the woman crying within. She was also talking rapidly.

"Help! Anyone! Help! This is the *Grimjoy!*" The crying broke up her words, but Goraksh knew anyone who heard her could still understand her. "We're being boarded by pirates! Help!"

The thought of the woman using the radio twisted Goraksh's insides into water. "She's calling for help."

"What are you going to do about it?" his father demanded. He released his hold on Goraksh's shirt.

"Help? Is anyone out there? There are pirates—"

Goraksh was unable to bear the thought of getting caught by the Indian navy or coastal patrols. No matter what, he had to stop the radio transmission.

THE CABIN WAS SMALL. A miniature galley and wet bar occupied the area to the left. A bed and shower cubicle occupied the forward and right sections.

A man, glassy-eyed in death, lay on the bed and rolled loosely with the pitch of the tethered yacht. He was in his thirties and looked American or European. Artificially blond hair was short and spiky. He'd been tall and fit, his skin bronzed by the sun. He wore brightly colored swimming trunks and was bare chested.

He'd also been dead long enough that his blood had settled in the lower part of his body. Goraksh had seen such things on television shows but he'd never seen anything like it in person.

The woman held on to the radio microphone as if it were a life preserver. She continued sending her message.

Goraksh shoved the pistol into her face as if he'd been doing it all his life. His finger wasn't even on the trigger.

"Get away from the radio," he shouted. Then he realized he hadn't spoken in English and that she probably didn't understand him. He repeated the order again as he reached for the microphone.

The woman jerked away. In the tight confines of the ship's cabin, she tripped and fell heavily. She had a death grip on the microphone and tore the unit from the wall in a shower of sparks.

As she floundered on the bed next to the dead man, she cursed Goraksh soundly. Goraksh didn't know what he was supposed to do next. He glanced back at his father as Rajiv came down the steps into the cabin.

Rajiv's eyes rounded in surprise.

When he swiveled back to look at the woman, Goraksh was stunned to see that she had a small black automatic pistol clasped in her hand. She continued cursing as her knuckle whitened on the trigger.

The detonation sounded loud in the cabin. Goraksh's ears ached with the blast and he was partially deafened. Sparks from the gun barrel singed his shirt. The bullet rushed in a heat streak beside his head, and he doubted that it missed him by more than an inch.

Goraksh pointed his pistol at the woman and fired back. He knew he'd missed, though. He'd hurried the shot and he'd missed. He barely even heard the reports because he was so scared. But there was more than one of them. He was sure about that.

Something burned into the side of his neck. He dodged away from it, but he knew he was already too late. He'd been shot.

The woman's head jerked violently. Her blood splattered the interior of the cabin and landed warmly on Goraksh's skin. He felt it ooze down his face as the woman fell over the dead man.

For a moment, Goraksh's knees wouldn't hold him. He thought he was going to fall. He tried to take a breath and couldn't. He wondered if he'd been shot in the throat. It would have been horrible to drown in his own blood.

Then his father was there. Rajiv slipped an arm under his shoulder and kept Goraksh on his feet. His father turned his head gently with the heated barrel of the .357 Magnum and surveyed the wound in his neck.

Goraksh felt his blood pulsing out of him. It soaked into his shirt. "Am I going to die?" he whispered.

"Not today," Rajiv replied in a choked voice. Tears glimmered in his eyes. "But I thought I had seen her kill you."

GORAKSH SAT in one of the upholstered chairs on the yacht's deck and watched his father's crew take the *Grimjoy* apart. They popped panels off the yacht and searched everywhere for hiding places.

So far they'd found a cache of money and a few weapons. There wasn't much of either. Whoever the yacht's owner had been, he'd known that if the vessel was searched those things would probably have been found.

A few of the pirate crew came over to congratulate Goraksh on killing the woman who'd almost killed him.

He didn't know what to say. He sat quietly and tried not to get sick thinking about it. Her face kept appearing in his mind, her mouth filled with curses until her head exploded.

Now that shock had worn off and his adrenaline had subsided, Goraksh felt the pain of his wound. His father had insisted on dressing it himself and had told him to keep pressure on it. The wound still bled through. His fingers came away stained with his blood. There was a lot of blood on his shirt, too.

He was convinced he should have been in a hospital. And he wasn't at all certain he still wasn't going to die.

One of the crew pulled himself up the ladder that hung over the back of the yacht. "Rajiv!" the man called. "Rajiv, come see what I have found!"

Instantly the crew swarmed the deck. Goraksh had to lean back in order to keep from getting bumped and jostled. He peered around Karam and spotted the pillow pack filled with a dirty brown substance.

Rajiv pushed to the front of his men and took the package. He slit a small hole in it with his knife, then dipped a finger inside. His finger came out coated with the substance. He smelled it but didn't put it into his mouth.

"Opium," Rajiv pronounced, and grinned. "How much have you found, Makhan?"

"A lot," Makhan said. Young and greedy, he was one of the newest additions to Rajiv's crew. "Maybe enough to make us all rich men."

The crew cheered. Karam even slapped Goraksh on the shoulder and then, realizing what he'd done, apologized profusely.

"Where did you find it?" Rajiv asked.

"In a blister glued to the hull of the ship. There's another on the other side." Makhan preened.

Goraksh knew about blisters. They were called that because they were fiberglass hulls designed to look like part of a speedboat or yacht. Smugglers used them to avoid cursory searches by law-enforcement personnel.

"Good. Let's get it out of there and onto the *Black Swan*," Rajiv said.

Unwilling to be trampled by the men as they hurried to get their new cargo aboard, Goraksh relinquished his seat. With nowhere else to go, he went down into the cabin.

There in the darkness, listening to the men as they cheered their good fortune, Goraksh stared at the dead woman. The bedding had turned dark with her blood. She'd been brain-dead from the instant his father's bullet had hit her, but her heart had continued pumping until nearly all her blood had emptied.

The man had died from a different cause. Rajiv had declared the man dead from a drug overdose. He'd had a pipe in his hand that had burned into his flesh.

Goraksh supposed that had been the impetus to get them started looking for illegal cargo and not just taking what they wanted from the yacht. There were always ship owners running opium out of India.

"What are you doing down here?"

Goraksh looked up at his father standing in the stairway. "I had to get out of the way."

Rajiv glared at the dead woman. "If you came down here to mourn the woman who shot you, then you're a fool," he spit.

"I didn't." But Goraksh felt guilty about her death.

"Then take her rings and jewelry. I had the men leave them for you. You earned them for facing the woman. If you feel like walking, then you feel like thieving. And if you wait too long, the body will become stiff and you'll have to cut off her fingers to claim them."

Without a word, Goraksh turned to commit himself to the task. He knew it would be hours before rigor mortis set in, and her flesh was drained of blood, which would thin her, too, but he didn't want his father hovering over him while he did the horrible deed.

Rajiv went back to concentrate on collecting the opium.

The woman had only a few pieces of jewelry but they all looked expensive. Evidently the man beside her had been both wealthy and generous.

Goraksh stuffed the rings, bracelet and necklace into his pocket. He looked down at the bed filled with death and wondered if that was how his life was going to end, too.

Then someone called out a warning on the upper deck.

Goraksh's wound throbbed as he ran up the stairs. The ship's crew crowded the railing on the starboard side. Most of them pointed at something in the distance.

A ship rolled out on the ocean as it rode the waves. When it started down the back side of one, the blue, gold and white Indian coast guard flag on the stern was revealed.

"Have we got all the opium?" Rajiv asked.

"Yes," Karam replied.

"Get aboard our ship. *Quickly.*" Rajiv glanced over his shoulder and spotted Goraksh. "Move."

Goraksh went as fast as he dared. He slipped on the

railing and nearly plunged into the water between the
boats. His father caught him and pulled him onto the ship.

"Make sail! Make sail!" Rajiv roared. He grabbed the
lines and unfurled the sails himself as he aided the crew.

The packets of cooked opium sat on the *Black
Swan*'s deck.

Terrified, Goraksh stood at the railing and watched
as the coast guard vessel sped toward them. There was
no doubt about the ship's destination.

The ship was an old one. India had put the coast
guard arm into effect in 1978 to aid the navy in patrol-
ling the nearly five-thousand-mile-long coastline. Still,
it possessed diesel engines and was making a good
speed. Goraksh was certain they would be overtaken
before they could get into international waters.

"Open the hold!" Rajiv yelled. "Quickly, quickly!"

The hold opened as the *Black Swan*'s sail filled with
air and the prow cut into the water. One of the crew
handed up an RPG-7. Rajiv took the rocket launcher and
ran to the stern.

Unable to believe what he was seeing, Goraksh followed. Surely his father didn't intend to attack the coast
guard ship. Even though it was one of the older craft in
service, it still had deck guns and her crew would be armed.

Instead, Rajiv took aim at the *Grimjoy* sitting dead
in the water. The rocket streaked across the water and
struck the yacht's engine compartment. Flames jetted up
and quickly spread.

Rajiv turned to one of the crew and took another
rocket from the utility bag the man carried. While the
yacht burned, he fit the second rocket to the launcher.

"We can't take the chance that we left someone alive on the yacht," Rajiv said. "Sail south toward international water. Thank the gods that we've got a strong wind."

Goraksh didn't think his father sounded as confident as usual. Tense and nauseous with fear, Goraksh watched as the coast guard ship altered its heading and steered for the burning yacht.

7

The four men in the speedboat concentrated on the dig members in the shallows near the cliff. All of the men had their backs to Annja as she treaded water and weighed her options. She had no doubt that violence was coming. She'd been around enough of it to recognize the signs.

Since she'd found the missing piece of the sword and claimed it as her own—or been claimed by it—she'd been thrown into a number of dangerous situations. She realized she was growing accustomed to it.

"I'm Professor Rai," Lochata announced.

"You're the boss?" the man with the rifle asked.

"I am."

"There's no man?"

"No."

The man took a cigar from behind one of his sun-burned ears, shoved it into his mouth and lit it with a lighter from his shorts' pocket. "Why would they put a woman in charge?"

Reluctantly, Annja released the gold coins she'd found and let them drop back to the sea floor. She marked the spot in her memory as best as she could and hoped that she would be able to find them again later. She dropped the snorkel, too, and swam toward the speedboat.

"I don't know," Lochata said.

"We're here to take your valuables," the man said.

"Valuables?" Lochata asked. "These are college students. They don't have money. *I* don't have any money. Nothing more than a few rupees."

"Then we'll take their computers and game systems," the man said. "Never seen college students without computers and game systems. Especially American college students." He waggled a forefinger. "And we know you have American students, old woman. We've heard about you."

"Then you also probably heard that we were hit by the tsunami last night. We lost everything," Lochata said.

"Don't lie to me," the man said.

"I'm not lying. We're waiting for help to arrive. The helicopter should be here at any time."

"If you're waiting for help, you're going to be waiting a while. A lot of villages were hit. They're not going to care about you." The man shifted his gaze among the college students. "You've got a lot of pretty girls." He pointed at one of the British women. "You. Come over here."

The young woman stepped back. Two of the young men, one of them Jason Kim, stepped in front of her to protect her.

The man's voice grew more threatening. "I told you to get over here." He worked the bolt action on the rifle. "If you don't, I'm going to shoot you through the head and move on to the next girl that catches my eye."

Annja reached the speedboat's starboard side. The craft sat low in the water now that it wasn't moving.

She caught hold of the boat's prow just in front of the windscreen. She thought about drawing the sword. It was there just within her reach. It would be awfully hard to explain, she thought. She decided not to pull the sword into the battle unless she had to. If she did need it, getting the weapon would only take a moment.

She pulled herself up and out of the water, then into a squat as the boat tilted toward her. Fear thrilled through her as she moved to stand. She was afraid of getting hurt, but mostly she was afraid of getting one of the students or Professor Rai hurt.

If you don't do anything, she told herself, they're going to get hurt anyway. These men are predators.

From the corner of her eye, Annja saw the man with the rifle turn toward her. Surprise turned his face slack, but he kept moving. Annja lashed out with a foot and caught the rifle as the man attempted to swing it around. The rifle roared and the bullet flew wide of its intended mark. Propelled by the kick, the rifle flew out of the man's hands.

That went better than I expected, she thought. Then she pushed herself to keep moving. The other men were in motion, as well.

Annja relaxed her arms and went down in a forward roll. The speedboat's hull felt blistering hot against her back for just a moment, then she was on her feet again.

The gunman still stood there, but he was reaching for a pistol tucked in the front of his shorts.

Fluidly, Annja spun into a side kick that caught the man in the face. The kick landed a little sooner than she'd expected, and pain shot up her leg.

Turn the foot, she chided herself. The impact should be against the ridge, not the flat of your foot, not the toes. You'll break the toes.

The kick had landed solidly enough, though. The man flew over the side of the boat as if he'd been fired out of a cannon. Before he hit in the water, Annja was already in motion.

The other three men swung their weapons in her direction. She dived over the windscreen and caught the man in the stern on the shoulders. She grabbed hold of the man's shirt and turned her body into a fulcrum as she swung down.

When she landed in a crouch behind the man on the gunwale, Annja hooked her hands under the man's arms, pulled him back toward her and threw him into the water, as well. He sailed over her as she swung around to face the remaining men.

She went on the offensive at once. A quick step to the right put her out of the lead man's sights and profiled him squarely in front of his partner. Annja took up a boxing stance but didn't want to risk further damaging her hands.

Instead, she threw Krav Maga elbow punches and knee kicks that collided with the man's face, groin and inner thighs. He couldn't defend against her attack and crumpled almost at once.

Even with the limited space, Annja executed a perfect

spinning back kick to his jaw. The man's head snapped sideways and he fell into the sea.

The last man had a pistol out. He tried to point it at Annja. She stepped closer and backhanded the man hard enough to stagger him. With a quick move, she plucked the pistol from his hand and threw it into the ocean. A backward blow to the temple sent him in a senseless sprawl to the deck.

Silence swept the shoreline as if the world had held its breath. Then the dig site members exploded with cheering and clapping.

Annja scooped up a pistol one of the men had dropped, checked the magazine and gazed out into the water.

The first man she'd thrown overboard had hold of the speedboat and was pulling himself up. Blood leaked from his nostrils.

"No," Annja said calmly as she aimed the pistol at him. "You're staying in the water until I'm ready for you to come out."

The man spit curses and threats.

Annja rapped him on the head with the gun butt, then grabbed a fistful of his hair and shoved him underwater. He fought to break her hold but couldn't. His fingers slid off her arm. She held him under long enough for him to realize how helpless he was.

When he stopped struggling, Annja brought the man's head and shoulders out of the sea. He grabbed hold of the boat's gunwale and coughed up water.

"Any more bad behavior," she told him, "and I won't pull you back up." It was an empty threat but the man didn't know that.

A quick glance back into the water showed that the other men she'd kicked into the sea still floated there with arms and legs outstretched. They weren't in any immediate danger of drowning.

Annja turned to the controls with the intention of using the maritime radio to signal for help.

One look at the torn wiring and the space where the radio had once been let her know that plan wasn't going to work.

"WHAT YOU DID back there was incredible," Lochata said.

Annja sat on a rock and finished lacing up her hiking boots. While she'd waded in the shallows she'd left her boots in the sun in hope that they'd dry out. They hadn't. She didn't like putting them on while they were still wet, but in case they had trouble with their four prisoners, she wanted protection for her feet from the bits of coral and sharp rocks.

Lochata handed her a bottle of water. "Skills like that generally aren't in the repertoire of an academic."

"I'm not your average academic," Annja said. She uncapped the bottle and drank. She felt parched, but she made herself drink slowly.

"Are you all right? During all the excitement I forgot to ask," the professor said.

"I'm fine." Annja's foot still hurt a little, but it was a slight and temporary pain. She knew that from previous experiences.

"Weren't you scared?"

"At the time, not so much. I didn't really have time to think about it. But now?" Annja held up a hand and let the professor witness her quivering fingers. Annoyed by her weakness, she flexed her hand and made a fist several times.

"Are you afraid now?"

Annja peered across the clearing at the four men seated on the shore with their backs to the cliffs. Their hands and feet had been secured with ordnance tape Annja had found in their tool kit aboard the boat.

"No, not really. But when everything slows down and you think of what could have happened, sometimes it can seem scary all over again. The main thing is no one got hurt."

"WANT SOMETHING to eat?"

Annja looked up from the sketch of the ocean seabed she was doing and saw Jason standing beside her. He held out a baked fish wrapped in a palm leaf.

"Sure. Thanks." Annja took the fish and found it almost too hot to hold.

"What are you going to do with your prisoners?" Jason asked.

Annja glanced over at the men.

The leader's face was swollen, and blood had crusted on his chin. Two of his companions slept, turned just enough to balance against the rocks at their backs. The fourth man sat in morose silence.

"We're going to turn them over to the coast guard," Annja replied.

Jason squatted beside her. Apprehension pulled at his mouth and eyes. "Do you think they meant it?"

"Meant what?" Annja was so involved in thinking

about the gold coins that she wasn't tracking the conversation well. She tried hard to remember what they'd looked like. All she could remember for certain was that they hadn't been of a uniform size.

"That part about killing us all when they got released," Jason said.

The prisoners had threatened to take revenge on everyone involved in the dig crew if they weren't released. Annja had offered to tape their mouths shut. Since the leader couldn't breathe out of his nose so well, he'd been the first to fall silent. The others had quickly followed his example.

"No." Annja pinched off a piece of fish and ate it. She savored the juices and was surprised at the flavor. "Is that garlic?"

Jason nodded. "Professor Rai found marjoram, thyme and wild onions. That stuff grows wild here. Guess that's why India was one of the main ports of call along the spice trade routes."

"Yes," Annja said, smiling.

"I was wondering if she knew what she was doing with the herbs. I thought maybe she was going to have us eating weeds. But she was right."

"Botany is a major part of archaeology. If you're going to study in this field, you have to know plants. Living and extinct."

"She said her mother had an herb garden."

Annja smiled. "That probably helped."

"I'm serious." Jason's voice was low, almost a whisper.

"About what?" Annja tried to remember what they'd been talking about.

"You know." Jason slid his eyes toward the prisoners. "Those guys."

Annja swallowed. "Those guys are going to have very full lives just dealing with all the charges against them. Since you're here, and I'm here, for that matter, the United States embassy is going to be involved. I think those guys are going to jail and not coming back for a very long time."

Jason nodded and seemed to relax a little. He rocked back on his heels and flopped onto the sand beside Annja.

"Would you have shot that guy?" he asked. "If he'd kept coming out of the water?"

Annja looked in the young man's eyes and tried to remember when she'd been that sheltered and innocent. Even growing up in the orphanage with all the nuns, she hadn't been totally protected. New Orleans, proud and predatory, had stood just outside the fence.

"What would you have done?" Annja asked softly.

Jason braced his forearms on his knees and shook his head. "I don't know. I hope I'd have been strong enough to shoot him."

"Not strong enough," Annja corrected. "Scared enough."

"Scared enough?" Jason said.

Annja nodded. "If you think you can handle a situation, you tend to build in second guesses. Should I do this or this? What happens if this goes wrong? What happens if I fail?" She was quiet for a moment. "Nature gave us a fight-or-flight response. Hardwired it right into our brains. When we're threatened, when we're scared enough, we simply react. In violent situations, I've learned that it's better to react than to think about reacting."

"You've *learned?*" Jason smiled nervously. "You've been involved in a lot of stuff like this?"

Annja thought about that for a moment and wondered how best to answer it. "Probably more than a lot of people."

"Do you get used to it?"

"No. Not really."

"You seemed like you knew exactly what you were doing."

"I trained in a gym to fight. That's not the same as what happened today."

"Why?" Jason asked.

"In the gym you know they're not going to kill you."

TWILIGHT SETTLED over the shoreline. Her work in her journal had totally absorbed her.

"Hey," someone called out. "There's something caught in the tide."

Annja looked up at the incoming tide. It was calm and peaceful. Of course, things had been pretty peaceful last night until the tsunami hit, she reminded herself.

She scanned the whitecaps as they rolled to the shore. The bonfire they'd lit for the coast guard to find them was bright. Silver moonlight turned the sea a dusky gray.

Flashlight beams played over the water for a moment, then focused on an object twisting slowly on the waves. Every now and again it crested the whitecaps. It resembled a manta ray.

But no manta ray Annja had ever seen boasted a hollow-eyed skull riding it.

8

Annja shoved her journal and pen into her backpack. She grabbed her Mini Maglite from one of the outside pockets and stood. The sand slid out from underfoot. She'd taken three steps toward the sea before she remembered the pistol she'd taken from the men.

She returned for it and tucked it into the back of her waistband. After taking the handgun and a couple of spare magazines for it, she'd thrown the other weapons into the sea. None of the other dig members knew how to use a pistol or rifle with any degree of skill. Even if they did, shooting someone was a whole different matter.

"That's a skull!" someone said.

"Man, you're imagining things."

"No, I swear. I saw a skull out there."

Most of the dig members scooted in toward the fire. Fear of the darkness and what lies in it was natural.

Lochata joined Annja. They walked down to the pounding surf together.

"Did you see it?" the professor asked quietly.

"Yes." Annja shone her light at the coiling mass riding in from the sea.

"Was it a skull?" The professor asked.

"It looked like a skull."

"Perhaps it was from the sacrificial pit."

For the professor to mention that, Annja knew Lochata was concerned that it came from somewhere else.

The mass mired up fifteen feet from the shore. It bobbed in the water, and every now and again the moonlight gave a hint of rounded ivory in the folds of whatever had floated in from the sea.

Annja stripped off her hiking boots and stepped into the water.

"Do you want company?" Lochata asked.

"No."

"Good. Let me know if you need me."

Annja shone the light on the mass that had wrapped around a rock in the shallows. Cold mud squished between her toes.

It's not an octopus, she told herself. It's not a manta ray. It's not anything predatory.

The folds moved on the tide and the skull rolled loosely. The complete revolution reminded Annja of the movie *The Exorcist*. Don't go there, she told herself. When was the last time you saw a possessed person? That doesn't happen.

The skull disappeared into the dark water.

Tense and ready to flee, Annja moved closer. She saw that it was some kind of mud-smeared fabric. She took hold of it gingerly and pulled while she tracked it with the flashlight.

The skull rolled out and fell into the water. Two arms, a handful of ribs, a pelvic girdle and a leg dropped out, as well.

Great, Annja thought.

"What is it?" Lochata called out.

"Some sort of cloth," Annja replied. "And most of a body."

"A body?"

"Yeah." Annja turned the flashlight so it struck the water surface at an angle. She used her peripheral vision to search for the ivory-colored bones beneath the gentle waves.

The first thing she found was a small gold box. A picture of a female *naga* was embossed on its side.

AFTER JASON DISCOVERED the body was nothing more than a collection of bones, he ventured out into the shallows to help Annja with her search. Between the two of them, they found the skeletal remains quickly. After weighting the fabric down with rocks, they piled the bones onto the cloth.

When they were sure they had all they were going to find, they pulled on the fabric and discovered it was rotted. Jason returned to shore for one of the coolers they'd found filled with bottled water and they put the bones in it.

SEATED BY THE FIRE, Annja examined the gold box she'd discovered.

"What is that?" the leader of the would-be robbers asked. "Is that gold?"

Too late, Annja realized that revealing what she'd found to the men could be a mistake. A glance at them showed they were all awake.

Annja ignored them and took her digital camera from her backpack. She took pictures using the flash and the light from the bonfire.

"Did you find a treasure out there?" the man asked again.

Without a word, Annja got up and walked to the pile of driftwood the dig crew had gathered. She took out a three-foot piece that wasn't quite as big around as her wrist.

She walked behind the leader and twisted his arm to get him to stand. When he was on his feet, she ran the stick between his elbows to pull his arms backward. Before he knew what was happening, Annja tripped him and sent him face-first onto the ground so that he was turned away from her working area.

She glared at the other prisoners. "Anybody else want to spend the night on his stomach?"

The three men looked away from her.

Annja felt bad about the rough way she'd treated the man. But they were out in the jungle away from the civilized world, and the man had found a way to terrorize the dig team even while restrained. She didn't owe him any kindness.

She quashed her guilt and returned to work.

A quick search through the artist's kit Annja kept for drawing sketches turned up a graphite stick. She turned to a blank page in her journal and did rubbings of all six sides of the cube. As she worked, water continued to drip from the object.

Several of the university students sat around her in a semicircle to watch her work. Their attention made her feel a little self-conscious. Jason had his own small group in eager attendance as he attempted to put the skeleton back together.

"Why is she doing rubbings if she's already taken pictures and drawn the cube?" one of the young men asked.

"Anybody want to answer that?" Lochata asked in her professor's voice.

No one did.

"Annja is employing a very old-school technique," Lochata said.

Thank you very much for the *old* part, Annja thought glumly. She was only four or five years older than most of the students and suddenly she felt ancient by comparison.

"Both the digital images and the drawings can be inexact to replicate the cube if Professor Creed so chooses," Lochata announced. "Neither of those methods offers a true measurement of the object. The rubbings allow her to be more exact."

"Why would she want to replicate it?" someone else asked.

"In case the original is lost," someone said.

"Or if she has to surrender it to the custody of the ASI after we're rescued," Lochata said.

That was exactly what Annja figured would happen. Lochata was required to report anything of material value to the Archaeological Survey of India when she found it. Especially since it had been found in boundary waters.

"It's leaking," one student observed.

"What does that mean?" Lochata asked.

That there's void somewhere inside the cube, Annja thought. She'd already been searching for a means of opening the cube to reveal its secrets.

Someone guessed correctly but Annja didn't know who it was.

As she turned the cube, she noticed that the *naga*'s eyes were mismatched. The snake woman looked almost cross-eyed.

She opened her Leatherman multi-tool and flipped open the smallest blade. Carefully, she turned the cube toward the best possible light and eased the blade's point against the left eye. Gold was a soft metal. She knew she could easily score it by mistake.

When she pressed on the left eye, nothing happened.

She switched to the other and pressed. The click of the mechanism releasing inside the cube was so slight she thought at first she'd imagined it. Then a line appeared at the top of the cube.

Silence filled the impromptu classroom.

After she returned the knife to her backpack, Annja twisted the top of the cube and found it turned easily. Two quarter turns freed the top. She turned the cube upside down and dumped the contents and the rest of the seawater into her palm.

The object inside the cube was a heavy gold ring. The ring was formed from the body of a female *naga* wrapping around herself to grab her tail in both hands above her head. Her breasts were turned outward and stood proudly. Sapphire chips glinted in her eye sockets.

"Okay," a guy with an American accent said, "somebody was into porn."

That got a laugh. One of the first things archaeology students learned in their course work was that sex wasn't a new invention. Nor was the incessant interest in it.

Annja reached for her camera.

"THE FIRST THING you gotta know is that you're not dealing with one body here. You're looking at two." Jason hunkered down on his haunches and smoothed his hair back with one hand. Two days in the wild without hair products had left him looking shaggy.

"How do you know that?" someone asked.

Annja surveyed the two collections of bones laid out on the sand. She already knew the reasons from the cursory examinations she'd given them.

"First off," Jason said, "these bones belong to a man and a woman." He took pride in his work, and Annja was glad to see he had a real, defined interest. He pointed to the pelvic girdle. "That belonged to a woman. Wider hips are always a giveaway. The skull belonged to a man." He picked up the skull and looked at Annja. "Can I keep this one?"

"No," Annja said.

"Spoilsport."

Annja gave him a half smile.

"How do you know that's a man's skull?" one of the women asked.

"Because the bones Jason's laid out with the skull are longer and thicker," Annja said to take away some of his thunder. After the question about the skull, he deserved to be taken down a peg or two. She touched the bump at the back of the skull. "Because the occipital protuberance is pronounced. And because the brow ridges are heavy."

Jason frowned at her. "You missed the teeth. Big teeth." He opened the skull's mouth to display the teeth. "Guys tend to have big teeth."

Annja nodded. "Big teeth."

A few of the men self-consciously touched their teeth.

"Who tore them apart?" the young woman asked.

"They weren't torn apart. The bones simply became disarticulated," Jason said.

"What's that mean?" one of the male students asked.

"It means that these skeletons were down there long enough for the flesh, cartilage and sinew to dissolve," Annja said.

She took pictures of the skeletons. How long were you down there? she wondered.

"What killed them?" someone asked.

Jason shook his head and cleaned his glasses on his shirt. "I don't know. I couldn't find any marks on them to indicate a violent death. At least, not on the pieces I have here." He grinned and rubbed his hands together. "Find me the rest of the bodies and maybe I can tell you."

"I'll pass on that."

"Who knows? By morning the sea might just wash the rest of the bodies in."

"Or maybe they'll just rise up out of the sea like those zombie pirates in that Johnny Depp movie," Jason suggested. He was immediately pelted with handfuls of sand that sent him into a hasty retreat.

"Great," one of the young women said. "Thanks for that, freak. I won't get any sleep tonight."

"Like you were gonna sleep anyway with murderers tied up only inches away from you," Jason said with a smirk.

9

Most of the dig crew slept eventually. There'd been nothing else to do. Lacking television and video games, and having only conversation for diversion after no sleep the previous night, they'd gradually passed out in the lean-tos they'd made from palm branches.

Annja and Lochata had shown them how to build the shelters toward evening when they'd given up on the coast guard arriving before night fell. Thankfully they'd found a few blankets tangled in the trees that had dried out during the day.

Lochata woke and looked around. When she saw Annja standing in the moonlight near the water, she got up, wrapped a blanket around herself to ward off the chill and joined her.

"Can't you sleep?" Lochata asked.

"Not yet," Annja admitted.

"Are you worried about those men?" Lochata nodded at the four bound men.

They lay by the fire close enough to be warm. All of them slept as if they hadn't a care in the world.

"Not really," Annja said.

"I could keep watch over them for a few hours while you slept," Lochata offered. "I don't know how to use the gun, but I could wake you."

"That's all right. I'm fine. Really. I'm used to staying awake a lot while I'm on a dig." Annja looked at the diminutive woman. "I guess your excavation didn't turn out as well as you'd hoped."

Lochata smiled. "At least now I know that all the research I'd done reaped a reward. The site was where I said it would be. It'll still be here if we come back next week or a year from now." She shrugged. "I can write the book with what I have now. This dig would just have provided more pictures to include. Today's readers like lots of pictures."

"I know." That was one of the reasons Annja took as many pictures as she did. She nodded at the bones lying on the sand. "I don't think those two are the only ones out there."

"Of course not. Ships have been lost out there for centuries."

Annja paused. She'd been thinking about whether or not she was going to tell Lochata what she'd suspected for hours. She didn't want to drag the professor away from her own work.

But in the end, what she thought had taken place out on the ocean was too exciting to keep to herself. More than that, she needed the professor's contacts. She'd have to get permission from the ASI if the ship lay in boundary waters.

"I think a ship is out there," Annja said softly. "I think the tsunami ripped something up from the bottom that's been buried a long time. That's why the fabric the skeletons were tangled up in was so well preserved. And there was something else." She quickly relayed her discovery of the gold coins before the speedboat had arrived with the four men.

"What ship?"

Annja shook her head. "I don't know. That's one of the things I want to find out. I've read your books. I know you've done shipwreck archaeology."

"Yes. Several times. This country is thousands of years old. Trade has gone on here nearly since the dawn of civilization. There are untold treasures—and I mean those of historic significance, not just monetary value—lying out there in those waters."

"I haven't done much marine archaeology, and I've never worked a shipwreck site before," Annja said.

Lochata looked at her. "You want to hunt for this ship."

Guilt stung Annja as she nodded. "It's just—I *know* it's out there."

"Of course you might follow your heart." Lochata smiled gently and patted her arm. "I appreciate everything you've done for these students. If not for you, some of them would have been lost to the tsunami. Others would have been lost today. I would be remiss to try to hold you."

"It's not just me," Annja said. "I want you to help me look for that ship."

Lochata hesitated. "We don't have the equipment to conduct such a search. We'd need a boat specially equipped with blowers to excavate a site."

"We might not need to excavate the site," Annja said. "If the tsunami ripped a shipwreck up from the sea floor, it might just be lying there."

"The boat would still be necessary to operate from."

"I know." Annja took a deep breath of the cool night air and let it out. "I might be able to call in a favor."

"A favor? That would be an awfully big favor," the professor said.

Annja silently agreed. She was thinking of Roux, the mysterious old man who had known about Joan of Arc's shattered sword and who claimed to have been alive for at least the past five hundred years.

"There's a man I've done some independent consultations for," Annja said.

"He must be a very rich man," Lochata said.

"He is." Annja had no way of knowing how much Roux was worth, but she suspected the old man had millions or billions put away. She knew that Garin did, too, and he pursued more money constantly.

"Then," Lochata said, "if you can make this happen, I would be happy to be part of your endeavor." She grinned widely. "You're not the only one who's been made curious by that *naga* statue."

HELICOPTER ROTOR BLADES woke Annja early the next morning. The sun was barely up and shadows of the night remained within the jungle.

She'd slept sitting up, her legs straight out before her,

ankles crossed. That had been a mistake as it turned out. Her legs had gone numb and she had trouble getting to her feet. When she could move she ran out onto the beach and waved her arms over her head.

By that time other people in the camp had risen and joined her. They stood and shouted at the blue-and-white-striped helicopter as it rolled out over the sea even though Annja knew their shouts couldn't be heard.

"They didn't see us," someone yelled.

Annja returned to her backpack and grabbed the flare gun she'd taken from the speedboat yesterday. She berated herself for forgetting, but she knew she'd been so tired that she couldn't be expected to remember everything.

A press of a button released the flare gun's barrel. She took out the night-use flare and shoved in one marked for day use. Without taking aim, she pointed in the helicopter's general direction and fired.

The flare gun bucked against her palm as the charge leaped from the oversize barrel and took flight. The flare screamed through the air and detonated. Red smoke spewed out across the sky.

Surely they saw that, Annja thought. They had to see that.

A moment later, the Indian coast guard helicopter turned around smartly and headed back. This time when they all waved, the helicopter pilot juked his craft from left to right to wave back.

"Get back," Lochata directed. She waved her arms to chase her students back into the jungle, away from the descending helicopter.

Annja stood in the tree line with her arm covering her

mouth and nose as much as possible so she wouldn't breathe in the dust and sand the rotorwash kicked up from the shoreline. Her sunglasses protected her eyes, but she felt the sand peppering her exposed flesh.

Once the helicopter was on the ground, three men in blue jumpsuits, light-gray helmets and life vests got out. They wore pistols in shoulder holsters.

"Professor Rai," one of the men called. He was young and intense looking.

"Here." Lochata stepped from the jungle. She joined the man and they started talking rapidly.

Another man stared at Annja and approached her. He gave her a quiet, friendly smile and moved slowly to take the flare gun from her.

"Just so no one will get hurt," the man said politely.

"I've got another gun." Annja pointed at the pistol by her backpack.

The man's face looked a little grimmer. "Is that the last one?"

"Yes."

With economy and grace, the man walked to the backpack and took the pistol. "This isn't something you'd ordinarily see an archaeologist carrying," he said.

"You haven't seen the Indiana Jones movies, have you?" Annja teased.

The man laughed. He had a nice smile and nice eyes. "Actually, I own all those movies on DVD."

"So do I," Annja said.

"I've also got a few of your DVD collections. *Chasing History's Monsters* is one of my favorite shows. I never miss it."

Terrific, Annja thought.

"I am Padhi, Miss Creed. It is a pleasure to make your acquaintance." He offered his hand.

Annja took his hand and shook it briefly.

"Now, would you like to tell me why those men are tied up?"

IT TOOK A LITTLE WHILE to get all the stories told, but the coast guard had brought supplies and passed those out while Lochata and Annja explained everything that had happened. In minutes breakfast—mostly protein bars and fresh fruit—was served.

"We can't transport you all back at the same time," Captain Yadav said when they'd finished. He was in his late thirties and already going gray at the temples. "But we can take a few of you now." He wiped at his stubbled face with a big hand. "Of course, we'll need to transport the prisoners, too. We can hardly leave them here with these young people."

"I heartily agree, Captain," Lochata said.

"We can take a few of you back to Kanyakumari. Afterward we can make arrangements to get the rest of you out of this place."

"I don't know that we'll be ready to leave," Lochata said. "Some of the students might wish to leave, but I'll have to ask them."

"Why would you stay?" the captain asked.

"If I can get my site reequipped, I plan on staying until I finish what I came here to do."

Captain Yadav nodded. "As you wish, Professor. Is there any way I can assist you?"

"I need to get word to the families of these students that they're fine."

"Of course. Get me a list of their names and their information. I'll see to it everyone is contacted even if I have to do it myself."

"And Professor Creed will need to be flown into Kanyakumari to make arrangements," Lochata said.

"Actually, her transportation has already been arranged."

Annja was confused. "*My* transportation?"

The captain nodded. "There was another search-and-rescue effort going on along with ours."

"Who was it?" Annja couldn't believe Doug Morrell would go to the expense of hiring a private team to find her.

The captain shook his head. "I only know that the man looking for you was known to my superiors. They asked me to give him all the information I could get as soon as we located you." He looked up in the sky as the sound of another helicopter grew louder. "Ah. There it is now."

Annja turned and looked. The helicopter looked like a small blob in the sky, but the rotors instantly identified it.

"Someone sent a search-and-rescue party after me?" Annja asked.

Yadav nodded. "Evidently you have influential friends."

10

The black helicopter with the corporate logo of a gold dragon in midflight landed out on the ocean on large pontoons. Annja watched as the helicopter sat there while a crew disembarked and waded through the sea to reach the narrow shore.

The new helicopter was larger than the coast guard's. The crew threw out anchors to hold the craft in position. They moved with crisp efficiency that spoke of military training. Without speaking, they spread out and secured the beach. All of them carried assault rifles.

A medium-sized black man with a shaved head and wraparound sunglasses approached Annja. He was smooth-shaven except for a soul patch on his chin.

"Ms. Creed." His voice was melodic, Caribbean by way of England, Annja thought.

"I am," Annja said.

"I'm Matthew Griggs, ma'am. I'm here on behalf of my employer to make certain you're safe."

"Who's your employer?"

"Mr. Garin Braden, ma'am." His face was like stone.

He looked familiar but Annja couldn't remember where she'd seen him.

"Do you have any ID, Mr. Griggs?" Annja asked.

"Yes, ma'am." Griggs reached into his pocket and pulled out a slim wallet. Inside was an official-looking license. He was a member of DragonTech Security Services. "Mr. Braden also told me to have you call him when I made contact."

"My phone's battery is dead," Annja said.

"Mine isn't, ma'am." Griggs handed over a slim phone.

"I don't have his number," Annja said.

"It's in the phone directory, ma'am."

Annja located the number and punched Talk. "I know he protects his private information," she said while the phone dialed. "I guess he wouldn't be happy if you lost this phone."

"No, ma'am. That number is a temporary one. After you call Mr. Braden, it won't be used ever again. I've memorized other numbers for him that I use."

The phone rang and rang.

"He's not answering." Annja handed the phone back to Griggs.

Griggs made no move to take the phone. "Perhaps we should give him a moment, ma'am. Mr. Braden was adamant about speaking with you."

Irritated and determined not to play Garin's little

game, especially since she didn't know the rules, Annja tossed the sat phone back to the security officer.

Griggs caught the phone in one hand without batting an eye.

"Can you fly me to Kanyakumari?" Annja asked.

"Yes, ma'am."

"How many others can you put on board that helicopter?" Annja knew some of the dig members wanted to go home. Some would help Lochata put together supplies to continue work at the site.

"Ma'am, I'm afraid that won't be possible. My orders are to take you, and only you, wherever you want to go."

"If you can't take anyone else, then I'll ride with the coast guard." Annja turned and walked away.

The phone chirped.

Annja never broke stride until Griggs called out to her.

"Mr. Braden would like to speak to you, ma'am," Griggs said.

If it hadn't been for the fact that the DragonTech helicopter could help ferry people, Annja wouldn't have taken the call. That was what she told herself, anyway. The truth of the matter was that she was curious as to why Garin had sent the security team.

She returned to Griggs and took the phone he offered. "Yes."

"Ah, Annja," Garin said. He sounded out of breath. "Good morning. I see you survived your debacle."

"I did." Annja listened to the noise coming from the other end of the connection. If she didn't know better, she would have sworn Garin was running through a forest with branches whipping at him. "What are you doing?"

As soon as she asked, she realized she didn't want to know. Garin Braden was a true sybarite. From what she knew of him, he loved women and luxury. Neither he nor Roux abstained from the pleasures of the flesh. She'd found Roux with women young enough to be his granddaughter on more than one occasion.

"Never mind," she said quickly when she envisioned what Garin might be preoccupied with. "I don't want to know."

She always found thoughts of Garin confusing. He was larger than life. His long black hair and goatee were always immaculate. He was a devilishly handsome man, but he also had a devil's heart.

"I was busy escaping my own debacle when you called," Garin said. "It was an inopportune moment to take a call."

"I'll bet." Was it a brunette, blond, or a red-haired moment? Annja wondered.

"I'm glad you survived," Garin said with such sincerity that she almost believed him. Fool, she chided herself. False sincerity and bare-faced lying were two of Garin's greatest abilities.

"Do you still have the sword?" he asked.

"Yes."

"Pity."

Annja smiled. As long as she had the sword, she knew Garin could never completely forget about her. The downside was that he would never forget about her.

"Why did you send the helicopter?" she asked in an effort to focus on another line of thinking.

"I thought you might need it."

"How did you know I was here?"

"It was on television."

"Television?" Annja couldn't believe it.

"Yes. On CNN and Fox News."

"How did they know?"

"Your television producer. That little cretin, Morrelli."

"Doug Morrell," Annja said.

"Whatever."

"CNN contacted Doug?"

"I think it was the other way around. It was a great performance," Garin said.

Annja felt as if the world had just opened up beneath her feet. "*What* was a great performance?"

"His tears and gnashing of teeth. I think he would have torn his clothing if he'd had time. This generation doesn't know how to properly grieve."

"Why would he be grieving?" Annja struggled to catch up. Events were twisting too quickly.

"Morrelli claimed that you were killed when the tsunami struck. I didn't believe that—"

Or you just wanted to confirm for yourself, Annja thought.

"—so I sent a security team I already had in the area," Garin finished.

"Thanks." Annja paced and glared out to sea.

"You sound angry," Garin said.

"I am."

"At me?" Garin sounded puzzled and a little hurt.

"No. At the cretin."

"Ah. Well, good. It's one thing to have you mad at me when I deserve it—"

"You deserve it," Annja said. "A lot."

Garin laughed. "That's debatable."

"Not with me." Annja turned and glared at Griggs. The man didn't wilt a millimeter. His sunglasses only reflected her and the beach. "Your officer here—"

"Griggs."

"Right. He tells me I'm the only one who gets to ride on the helicopter."

"You want to take someone?"

"Several someones. There are a lot of scared college kids out here. I'd like to see that some of them get home."

"Let me have a word with Griggs."

Annja gave Griggs the phone.

"Yes, sir," he said twice, and handed the phone back to her.

"Take as many as you can fit onto the helicopter," Garin said.

"Thank you. Can I keep the phone for a while, too? Mine is dead."

"Of course." Garin took a breath. "I hate to arrange an escape-and-run, but that's how these things are generally best handled. And I've got an escape of my own to finish."

"Good luck with that," Annja said.

"Luck has nothing to do with it," Garin declared. "I'm a master planner. This little—"

He was cut off by a burst of machine-gun fire. He cursed loudly and the phone went dead.

Annja tried to call back, but the phone only rang a couple times and went to an answering service.

"Problem, ma'am?" Griggs asked.

"Garin got cut off," Annja replied. She was surprised how much the idea of him hurt somewhere worried at her.

"Satellite signals sometimes get—"

"By machine-gun fire," Annja interrupted.

"Ah." Griggs nodded and took a deep breath. "Well, ma'am, I wouldn't worry too much. Mr. Braden has been taking care of himself for a long time. I've found him to be quite capable."

"I'm not worried," Annja said. But she knew she was lying.

"Of course you aren't, ma'am."

Evidently Griggs knew she was lying, too.

11

James Fleet arrived in the Kanyakumari morgue two minutes ahead of the appointed time. He descended the dark, narrow stairs that led to the basement. Despite his familiarity with the dead, he'd never truly learned to relax around them. Dread filled him at what he knew he was about to see.

He knocked on the door.

"Come in," a voice called from within.

Fleet opened the door and entered.

The morgue was larger than he'd expected. Four tables lined the wall on the right. Sheets covered the newly dead, and Fleet was glad he didn't have to see them. Stainless-steel vaults lined the wall on the left. Cabinets with surgical instruments and chemicals occupied the wall ahead.

A thin old man in a white coat and a turban worked on the body of a middle-aged man with three bullet holes in his chest.

"Who are you?" the old man asked.

"Dr. Singh?" Fleet asked because he didn't want to explain himself only to find out he was talking to the wrong person and have to explain everything all over again.

"Yes." The man paused and leaned on the surgical table as if he were tired. But he was wary.

Fleet was six feet tall and weighed close to two hundred pounds. With sandy hair and freckles across the bridge of his nose, he'd never pass for a local. His fair skin was still pink from the sun exposure he'd had yesterday. He blamed his mother's Welsh ancestors for that. He wore slacks, a knit shirt and a loose windbreaker to cover the 9 mm pistol in the holster at the small of his back.

He showed his identification to Singh.

Singh peeled off his bloody gloves, dropped them into a biohazard container and took the identification. He swapped one pair of glasses for another.

"Special Agent James Fleet of the International Maritime Bureau," Singh read aloud. He squinted up at Fleet. "You're British."

"Does that make a difference?"

"No. Just an observation. Usually the IMB allows the coast guard or the Indian navy to handle investigations like this."

"Special case," Fleet said.

Singh touched his own nose. "You should be careful out in the sun."

"I know." Fleet took the ID back. "My office called to let you know I was coming."

Singh nodded. "They did. What may I do for you, Agent Fleet? Your office didn't tell me what it was you wanted."

"You can just call me Fleet."

"Of course." Singh waited patiently.

Fleet put his ID away and took a folded printout from his pocket. "You received a gunshot victim yesterday."

Singh held up four fingers. "I received four. Perhaps you would like to clarify."

"A woman." Fleet unfolded the paper displaying the image of the woman who'd been shot in the head with the .357 Magnum.

"Doug Morrell."

"You sound really chipper for somebody who's supposed to be grieving over me," Annja declared as she sat in the passenger seat of the DragonTech Security helicopter. The luxurious cabin was soundproofed enough that she could make the phone call without being drowned out by the rotorwash.

There was a pause. Then Doug's voice turned angry. "Who is this? How dare you call in and claim to be Annja Creed, may God rest her soul?"

Annja counted to three. It was the best she could manage. "Doug, it's me."

Anguish—fake and syrupy to the ears of someone who truly knew him—filled his words. "Oh, God. You sick, pathetic person. Annja Creed was one of the best friends I've ever had."

"I'm. Not. Dead," Annja stated evenly. The fierce tone in her voice drew the attention of the archaeology students seated around her. A few of the security team's heads turned, as well.

"Why don't you come down here and tear the heart right out of my chest?" Doug wailed. "And stomp on it?"

Annja didn't know what to say to that.

"Are you close enough to come down here?" Doug asked.

"I'm still in India."

"Good," Doug whispered. "Stay there. Stay out of sight."

"What?"

"I'll call you in a few minutes." Doug hung up.

Not believing what had just happened, Annja took the phone from her face and stared at it. She closed it and tried Garin's number again.

"Hello," Garin answered.

Annja was surprised to discover how relieved she was. The sword bound them and she knew as long as Roux, Garin and she lived that they would have some connection to each other.

"You're still alive." She tried to sound neutral, as if they were discussing the weather.

"I am. Are you happy about it? Or disappointed?"

"Disappointed."

"You don't lie very well."

"I lie just fine." Annja bristled at his confidence. She leaned back in the executive chair and stared down at the jungle below. They hugged the coastline as they flew toward Kanyakumari.

"Maybe to someone else. Why did you call?"

The question caught Annja by surprise. She'd called

because she wanted to know how he was, but she didn't want him to know that.

"I didn't know if I'd thanked you for the use of the helicopter," she said weakly.

"You did. And thank you for calling to check on me."

"I didn't call to check on you."

"Have you called the cretin?"

"I have. He accused me of pretending to be me."

"Of course he did."

Annja was at a loss. "Why would he do that?"

"Because, at the moment, it's more profitable for the television show if you're dead."

"Why?"

"Because they're advertising the *Chasing History's Monsters: The Annja Creed Memorial Collection*."

"What?" Annja was stunned.

Garin chuckled. "I have to admit, I was really surprised at how quickly they pulled it together. Do you have your computer?"

"Yes."

"Look at the show's Web site. I'll talk to you later. I've still got a few things to handle here."

"Where's here?" Annja asked before she could stop herself.

"Not there," Garin assured her. He hung up.

Annja cursed Garin's arrogance as she took her computer from her backpack. She'd plugged it into one of the outlets built into the helicopter to charge the battery. When she started to attach the minisatellite receiver through the USB slot, Griggs addressed her.

"You don't need to use that," the security officer said. "The helicopter has wireless capability built into it."

Annja went online and accessed the *Chasing History's Monsters* Web page.

Normally the page opened up on a side-by-side profile of Kristie Chatham and Annja and had a current Top 40 song playing in the background. Now the focus was all Annja.

A black border ran around the page. The words "Annja Creed RIP" scrolled across the page. She gazed at the information posted about her. Most of it was incorrect or wildly exaggerated. The birth year was seven years earlier than it should have been.

"Hey," Jason said, peering over her shoulder, "you're dead." He blinked. "And you're older than I thought you were."

"Jason," Annja said as calmly as she could.

"Yeah?"

"That's a cargo door behind you. One more word and I'm going to drop you through it."

Jason mimed zipping his lips, turning a key and throwing it away.

Just then the page split and a pop-up flashed onto the screen.

Chasing History's Monsters presents the memorial DVD of Annja Creed's most memorable moments. Includes bonus footage of Kristie Chatham's eulogy of her beloved cohostess.

Video footage of Kristie followed immediately. She dabbed at her eyes with a handkerchief, and her movements only emphasized her cleavage.

Another video ran. This one showed Doug Morrell sitting at his desk looking devastated. A single tear leaked down his cheek as he spoke.

Annja didn't have the volume turned on so she couldn't hear him. She was pretty sure she didn't want to.

Doug finally buried his head in his arms in a theatrical gesture.

Unable to bear more, Annja shut the Web page down. Focus, she told herself. You've got work to do.

She reached for her bag and took out her camera SDRAM card and got ready to try tracking down the statue of the female *naga*.

One thing she'd discovered about herself was that whenever things got out of her control, she could focus on work. Working always helped her get through.

"HAVE YOU IDENTIFIED HER?" Fleet asked.

Dr. Singh pulled open one of the stainless-steel vaults. Inside, the woman's body lay in a plastic container that obscenely reminded Fleet of a sandwich bag.

"Not yet," Singh said.

"You've got pictures of her?"

Singh nodded. "I can give you a file containing what we have on her. It isn't much."

"That would be most helpful," Fleet said. "You haven't done an autopsy on her?"

"There was no need. Her death is simple to figure out. She was shot in the head with an extremely powerful handgun."

"A .357 Magnum."

"Exactly."

Fleet looked at the nightmarish head wound and felt badly for the woman. She was in her mid-thirties, about his age.

"You came here because of the bullet." Singh smiled in sudden understanding.

Fleet looked at the man. "Yes. The gun has been used before in murders. All of those murders tie to a particularly vicious pirate who we know has been operating in the Indian Ocean."

"Several pirates operate in the Indian Ocean," Singh said.

"Eventually," Fleet replied with a small grin, "I'll get them all. Today I want the one who killed this woman."

"I ran a toxicology report on the woman," Singh said. "At the time of her death, she was under the influence of opium. The coast guard officer I talked to said that drug paraphernalia had been found on the boat where she was killed."

"The report also said a man was brought in with her."

Singh rolled the vault back into the wall and picked another unit. He pulled it out and revealed the dead man inside the plastic bag.

"I don't see any marks on this one," Fleet said.

"There aren't any. He died of a drug overdose."

"Opium?"

Singh nodded.

"Was he a dealer?"

Singh shrugged. "Until he came in on a gurney with a toe tag, I'd never seen him before. I have no way of knowing that. Perhaps the police can help you."

Fleet carefully took a deep breath. The stink of the morgue was making him ill.

"I haven't been here before," Fleet said. "Maybe you could direct me to the police station."

12

Only a few minutes out of Kanyakumari, according to the helicopter pilot, Annja posted the images of the female *nagas* on her favorite archaeology Web sites. She chose the images that showed the egg-shaped *naga* opened and closed, then did the same for the cube. She also posted images of the *naga* ring.

> I found these on a recent trip to India. I was on one of the beaches in the Tamil Nadu District. I turned them over to the Archaeological Survey of India, but I'm curious about their history. Can anyone help?

She'd just finished checking to make certain the postings had cycled through and were up when her phone rang. Her phone sat on the floor of the helicopter cabin where she'd found the outlet and plugged in the charger.

A glance at caller ID showed that it was a New York

number, but it wasn't one that Annja recognized. She flipped the phone open and said, "Hello."

"Annja," Doug gasped. "Thank God you're alive. I mean, when I heard the story I couldn't even begin to imagine that you'd been killed."

The irritation she felt at Doug returned full force. "That's not exactly what you were saying a few moments ago."

"I couldn't let anyone in production know you were alive."

"Why?"

"Because—" Doug fumbled for a reason. "Because they might try to kill you again."

"Who? The people in production?" Annja asked.

"No. Not production. You know. The people who tried to kill you in India."

"Doug," Annja said slowly, "no one tried to kill me." She stopped herself. Technically that was incorrect. "Well, someone did try to kill me, but you were evidently telling everybody I died in the tsunami, which was before people tried to kill me."

"That's right!" Doug exclaimed. "You died in the tsunami attack!"

"Doug."

"Yeah?"

"I'm not dead. And tsunamis don't attack," Annja said, although, admittedly it hadn't felt like that the night before.

"Oh, yeah. I knew that." Doug sighed. "It's just that right now isn't a really good time for you to be alive."

"Being alive works for *me*," Annja said.

"I know. I know. And it's going to work really well

in the ratings when we let everybody know you escaped that tsunami's clutches."

Annja started to point out that tsunamis didn't have clutches, then decided against it.

"But we've got this *thing* going on right now," Doug said. "It's kind of important."

"You mean the Annja Creed memorial DVD?"

"Oh. You know about that?"

"It's being advertised on the show's Web site."

"You *never* look at that Web site," Doug complained. "Why are you looking at it now?"

"Someone told me you were telling people I was dead," Annja said.

"The story was on the Internet. I didn't start it. Somebody must have known you were in India doing whatever it is you're doing."

"Looking for space aliens," Annja said.

"Get out of here!" Doug exploded. "If you're doing something like that, why aren't we covering it? Aliens do all those cow mutilations and stuff. They'd be great to have on the show."

Annja put her free hand to her head and rubbed her temples. Talking to Doug sometimes gave her a headache.

"I'm not looking for space aliens," she said. "I told you I was here—"

"Digging up human-sacrifice victims. See? I remember. I do listen."

"Good. Then listen to this— I'm alive."

"Um. Yeah. About that." Doug took a breath. "I was thinking—"

"Doug, I don't want to be dead," Annja shouted

"Hey, you're getting a great cut of the memorial DVD."

"Dead people can't use their credit cards," Annja said.

"I don't think they know you're dead in India."

"The story ran on CNN. They probably know I'm dead in India."

"I didn't know they got CNN there."

Annja sighed.

"Anyway," Doug said, "I thought you were camping in a tent."

"I was. Tonight I'm not. Tonight I'm staying in a hotel with a bed and a bath. I'm going to pay for it with my credit card."

"What about the camping-out-and-getting-close-to-nature thing?"

"You don't go on an archaeological dig hoping to get close to nature," Annja said. "It's the workload of eighteen- and nineteen-hour days and a long trip to the dig site that necessitate staying at a base camp."

"Sounds fun."

"It isn't. All kinds of insects crawl into your sleeping bag in a cot that's about half the size you need. I want a hotel tonight, Doug."

"The DVD is selling *really* well," Doug pleaded.

"And a bathtub big enough to swim in," Annja said.

Doug was silent for a moment. "You know, sometimes you're hard to get along with."

"Me? I didn't declare you dead."

"I mourned you on national television. My mother says I was wonderful."

"Terrific."

"It wouldn't hurt you to do this one little thing for me."

"Being dead," Annja said, conscious of all the university students watching her, "is not a little thing.'"

"I'll make you a deal," Doug said.

"I don't like doing deals with you. Every time I do, I feel like I get hosed."

"You're killing me here. If I go up and tell marketing that you're alive, they're going to freak. I'm freaking just thinking about telling them."

Annja sighed.

"Let me pay for your hotel room," Doug offered. "You stay dead a few more days, and I'll pick up your hotel tab. And room service and cable."

"What if the hotel has seen the news reports that says I'm dead?" Annja asked.

"I won't reserve the room in your name. I'll reserve it in someone else's."

Annja glanced outside the window and saw the airport coming into view. She didn't have time to argue. "Get me a hotel room, Doug."

"Cool! Fantastic! I'll get you a suite!"

"How long do I have to stay dead?"

Doug was quiet for a moment. "It's three days until the weekend. Let us have three business days to sell the memorial DVD and you can resurrect on Saturday. Or Sunday if you think that's more fitting."

"I don't like doing this," Annja said.

"I kind of got that."

"Besides the room, I want to talk to you about a boat."

"A boat?"

"If I'm going to stay dead a while longer, I'm going to need a boat."

"For what? A Viking funeral?"

"No." Annja hesitated, then quietly went on. "I think I've found a shipwreck. I'll need the boat to look for the shipwreck."

Doug sounded wary. "Is this boat going to cost a lot?"

"Probably," Annja said.

"I can't promise a boat."

"I can't promise I'll stay dead," Annja warned.

"That's blackmail."

"You started it."

"Can we get a story out of this?"

"Maybe. It's a shipwreck. If I find it, there's going to be a story," Annja said.

"Any monsters?"

Annja thought for a moment, then said, "A *naga*."

"Another what?"

"Not *another*. A *naga*."

"Still clueless," Doug said.

"A creature that's half female, half snake," Annja explained.

"Which half is which?"

"Female from the waist up."

"Okay. That sounds good. Sounds photogenic," Doug said.

Annja leaned back in the seat and tried to get comfortable. It didn't work. She couldn't believe how crazy her life—her *death*—was getting.

"Can you get me any pictures of a *naga?*" Doug pressed. "I want to run it by the number crunchers in marketing. Pictures will help. They don't listen very well, and you know they're not very imaginative."

Annja truly believed the lack of listening was a job requirement among the show's production crew. "I've got pictures. I'll e-mail you," she said.

"Set up a new account and e-mail me with any questions you have. We've got to keep you dead for as long as we can."

Annja sighed. "Sure. Just look for a posting from DeadinIndia. And I want that boat."

13

Fleet stepped into the police station from the humid heat of the streets. The day hadn't even fully started and the climate was already almost unbearable.

As soon as he was inside the building, Fleet felt at home in the boiling clutter and sense of urgency. The police station held an array of desks and computers with attendant police officers and inspectors.

Fleet gave his name at the desk and was shown to a chair. He waited for five minutes and filled the time by staring at the bulletin board on the wall across the hall.

He didn't see what was on the bulletin board, though. All he saw were the faces of the dead left by the pirate with the .357 Magnum. Those faces were broken, shattered, charred and vacant. Fleet had been tracking the path of destruction for fourteen months. The bullet the coast guard found on the burning yacht had been a fortunate thing.

And you're the closest you've ever been to him, Fleet reminded himself. So bloody well watch your step and go slow. Don't lose this.

"Special Agent Fleet."

When he glanced up, Fleet saw a short, thin Indian man in a wrinkled suit standing nearby. Probably in his late forties, the man had iron-gray hair, a round face and a neatly trimmed mustache.

"I'm Fleet."

The man offered his hand and spoke in a clipped British accent more polished than Fleet's own. "Inspector Phoolan Ranga at your service."

"Pleasure to meet you," Fleet said.

"Likewise, I'm sure." Ranga appeared to be a man who got down to brass tacks quickly. He tapped the file Fleet had in his hand. "You've been by the medical examiner's office, I see."

"I have. I like to get an early start on the day when I can."

"As do I. I've not been home since yesterday." Ranga looked at the controlled chaos going on in the room.

"If this isn't a good time for you—" Fleet said, though he was loath to lose the opportunity to speak to the man. There was no telling when another meeting might be arranged.

"There's never a good time, Special Agent Fleet."

"Just Fleet, please." Fleet didn't like being addressed by his title. It was a holdover from his days with the special boat service. He'd been chasing terrorists in those days, not pirates.

"Of course." Ranga smiled again. "I know a pushcart where we might get a decent cup of tea if you'd like."

"I'd like to see your files." Fleet wanted to remain focused on his task.

"I know those files," Ranga said. "We can talk. You can tell me what you know. I can tell you what I know. And I'll make sure you get copies of those files delivered to your hotel room."

Fleet felt certain there was no way to convince the man otherwise. "All right."

Out in the street, Ranga waved to the right toward a vendor.

Fleet walked with the man but kept his senses alert. During the eleven years he'd spent with the special boat service as a part of the Royal Marines, he'd been in many rough port cities that were all too ready to take advantage of strangers.

The sounds of the city were strange to Fleet even though he'd been there a handful of times. Before, he'd only stayed long enough to investigate what was left after the pirate attacks.

"After your office called and I agreed to meet with you," Ranga said, "I took the liberty of looking you up. I hope you don't mind."

It's a little bloody late now, isn't it? Fleet thought unkindly. But he said, "No. I don't mind." He would have looked him up, too. In fact, he'd looked Ranga up and had discovered nothing but good things.

"You were with the special boat service," Ranga said.

"I was."

"You look young."

"Thirty-four. Not as young as some."

Ranga grinned. "I could be your father."

That surprised Fleet. The man didn't look that old, and he hadn't been looking for an age when he'd researched Ranga.

"I have a daughter a year older than you. She's made me a grandfather twice over."

"Congratulations."

"Thank you. You were a commander," Ranga said.

"I was." Fleet struggled not to sound bitter.

"Weren't you young to be a commander?"

"Some thought so. I got the job done, and that's what my superiors were looking for—someone who could get the job done."

Ranga nodded. He stopped at the pushcart and ordered two cups of tea. He took his plain. Fleet added milk because he knew the tea would be strong.

"What did you do in your military career?" Ranga asked.

Fleet tried not to let his irritation get the better of him. He wanted to talk about the case, not his military background. But he knew the Kanyakumari police didn't even have to talk to him. As a special agent for the International Maritime Bureau, he was strictly civilian and dependent on the goodwill of the law-enforcement agencies he coordinated with.

"I patrolled coasts and tracked down terrorists," Fleet answered.

"Sounds exciting."

"Some days."

"Young men like excitement." Ranga walked

through the neighborhood. "You were involved in several altercations."

"One too many," Fleet agreed.

Ranga looked at him.

"You're wondering why I'm not still with the special boat service," Fleet said.

"It had crossed my mind."

"I was honorably discharged after my last mission. My team and I were ambushed by terrorist forces running drugs from Turkey to Bristol. They'd made us. We didn't know that until it was too late." Fleet looked away, no longer focusing on the Kanyakumari street, but back on that fast attack boat the night the world had blown up in his face.

"Two men on my team were killed. I caught shrapnel in my right eye that blinded me." Saying it now was almost easy. He touched the smooth lines of the scar on his cheekbone. The surgeons had been forced to rebuild it, as well. "They matched the prosthesis to my left eye. It's hard to tell the difference if you don't know what to look for."

Ranga shook his head. "I didn't mean to pry."

"I was also hit in my left shin by a .50-caliber round. The damage was too severe and the doctors had to remove the lower third of my leg."

"I couldn't tell."

A little bit of pride mixed in with the distant hurt that remained from all he'd lost. He'd worked hard to smooth out his gait. "The transtibial prosthesis they have now is good stuff. I was lucky because they saved my knee. Transtibial amputees can still compete in a lot of sports. But the Royal Marines decided that they couldn't use a one-eyed, one-legged commander. I was given a decent pension that I tried to drink up every month for a while. Then, when the drinking got old, I realized I had to get on with my life."

"So you joined the IMB?"

"It wasn't easy. Turns out the IMB didn't want a one-eyed, one-legged special agent in the field, either."

"Yet here you are," Ranga said.

Fleet nodded. "I'm nothing if not determined."

"I apologize for asking."

"Don't. I would have done the same thing if I'd seen someone like me walk into my office," Fleet said.

"Have you had breakfast yet?"

"No."

"Then permit me to buy you breakfast and we'll talk about Rajiv Shivaji," Ranga said.

"YOU PULLED A HIT on the .357 Magnum bullet, as well," Fleet said. He sat across the table from Ranga in a modest diner only a few blocks down the street from the police station.

Outside the big window, a man rode an elephant down the street. A large, vibrantly colored blanket covered the elephant's back and sides. The creature's trunk waved dispiritedly.

"I did." Ranga broke off a piece of chapati and dipped it into a dish of coconut chutney.

Fleet ate mechanically. "How many hits did you get?" he asked.

"Three. All confirmed."

"All here in Kanyakumari?"

Ranga nodded and sipped his tea.

"We matched the bullet that was recovered from the yacht with five pirate attacks," Fleet said.

"Whoever owns the gun faces serious trouble," Ranga said.

"You mentioned Rajiv Shivaji's name," Fleet pointed out.

"I did."

Fleet waited as patiently as he could.

"Unfortunately," Ranga sighed, "that particular gun isn't registered to anyone."

That caught Fleet off guard. "Then how did you come up with Shivaji's name?"

"Because I talked to a man who said he sold it to Rajiv Shivaji four years ago."

Anticipation flared in Fleet's belly. "He can tie Shivaji to the weapon?"

"He could have. I had the conversation with the man only six days ago. The next day he was dead. He was knifed in a fight that broke out in jail."

"Do you think Shivaji knew he'd talked to you?"

Ranga shook his head. "I think it was just bad luck." He paused to break off another piece of chapati. "But it was bad luck all the way around. If I hadn't talked to this man, hadn't been able to connect him with a burglary that took place six years ago, I wouldn't have known Shivaji had purchased the gun from him."

Fleet sipped his tea and thought about that. "If Shivaji knew you'd made the pistol, he'd have gotten rid of it. He definitely wouldn't have used it on the attack on the yacht yesterday."

"Perhaps. You're thinking too much with an organized mind. That is something you or I would have done had we known. Pirates are often superstitious. Shivaji believes the pistol has been blessed," Ranga said.

"Blessed?"

"By a holy man."

Despite the situation, Fleet couldn't help smiling in disbelief. "Why would a holy man bless a pistol?"

"Because the man who carried it was doing so to protect him."

"And Shivaji believes that blessing extends to pirates?"

"Perhaps it does. The man I talked to was silent for four years about his transaction. Now he's dead and we have no testimony to offer in a court of law to go after Shivaji."

"The blessing hasn't been entirely effective," Fleet said. "Now we know about Shivaji."

14

An elephant blocked the street ahead of Annja. She stopped and watched the creature for a moment. It wasn't an everyday sight in Brooklyn, though that city had its own surprises for out-of-towners.

The man on the elephant conversed with a man driving a taxi. While they were talking the elephant lifted its tail and pooped.

It was an impressive feat, almost like a concrete truck off-loading to fill up a concrete form. Annja was willing to bet small children and livestock could be lost in the mass that resulted.

Tourists instantly scattered to the other side of the street. A shopkeeper with a broom burst out of a nearby building and started screaming as if he'd been mortally wounded. He pointed to the steaming pile of elephant dung that littered the street and the sidewalk in front of his shop.

Tourism's not always as great as they make it look

on the travel shows, Annja thought with a smile. She walked to the other side of the street, as well.

The man on top of the elephant yelled back at the shopkeeper. Then the shopkeeper whacked the elephant with the broom several times. The elephant rider protested vigorously and shook his fist. Finally the broom broke and the elephant lurched into motion. The two men continued to yell threats at each other.

Annja's phone rang as the altercation ended.

"So the reports of your death *are* greatly exaggerated," Roux said when she answered.

Annja recognized the old man's raspy voice at once.

"I thought about ordering one of those memorial DVDs your television show is putting out," Roux went on. "Especially since Kristie Chatham is offering a calendar with the package."

Of course, beneath the old-world charm he was snarky to the bone, Annja thought. "I'm surprised you noticed," she said.

"That they were offering the calendar?"

"That I was alive."

"Generally I notice most when you're pestering me with questions."

"You called, not me," Annja said.

"I thought I was returning your phone call," Roux replied.

Annja thought back. She *had* called. As soon as she'd gotten off the helicopter at the airport and caught a taxi, she'd called. "You are returning my call," she admitted.

"I assume you just phoned to let me know you were alive."

"I thought you might be worried."

"Why?" Roux asked.

Pain and anger warred within Annja. She liked Roux. There was something inherently good about the man, just as there was something inherently evil about Garin. Roux was smart and strong, and he'd lived through more history than she was ever going to see. I don't know why I look to him for approval, she thought. But I do.

"I—" She hesitated. "I just thought you might be concerned."

"I never worry about you, Annja," Roux said.

How was she supposed to take that? she wondered. She was sorry she'd ever placed the call.

"You have Joan's sword."

My sword, she thought, but didn't say it.

"It will get you into danger from time to time. Though I've never heard of it attracting a tsunami. However, it wouldn't have come to you if you weren't clever enough to see your way through trouble. But seeing as how you're on the phone, how are you?" Roux asked.

"I'm fine." Annja kept walking toward the public piers. She'd been told there was a man who had a boat for hire that could serve as a base of operations to look for the shipwreck.

"I know you were there with a group of students," Roux said.

That surprised Annja. She'd talked with Roux briefly before leaving Brooklyn. She worked to stay in touch with him even though he only called when he wanted her to research something for him. Most of the time their relationship seemed like a one-way street.

"Are the students all right?" Roux asked.

"Yes. We were lucky." Twice, Annja added silently.

"That's good. When are you heading home?"

"Soon," Annja answered. And that made her a little angry, too. He'd asked as if it would make some sort of difference to him. And she knew it wouldn't.

"Well, then, keep safe," Roux said.

"Do you have a minute?" Annja asked.

Roux hesitated. "Perhaps one or two."

"Roux?" a young woman's voice called out. "Are you coming back?"

"That's Julia," Roux said. "She's my…masseuse."

"Right," Annja said. "I think I found a shipwreck."

"Good for you. There are a lot of them down there. I'm surprised more people aren't finding them."

"This one's weird," Annja said.

"How so?"

"I've got two items posted on the archaeology sites I resource. I thought maybe you could take a look at them."

"Why?"

"Because they're really old. You're really old. If you knew something, I thought maybe you'd pass it on."

Roux growled testily. "Just because I've lived a long time doesn't mean I've seen everything."

"Then maybe you'd like to look at something you haven't seen before."

"What is it?"

"You know what *nagas* are?" Annja asked.

"Of course. Who do you think created them?"

The question surprised Annja so much that she stumbled for a moment. "You?"

"No." Roux sighed. "You are so gullible."

"Not gullible," Annja argued defensively. She hated it when Roux got the better of her. But he'd done so much, seen so much, that almost anything seemed possible where he was concerned. "I'm tired. I've been up for two days."

"Then you should get to bed."

"I'm going to as soon as I get to the hotel. Look at the images. If you think of something I might like to know, call me or e-mail me," Annja said.

"Okay." Roux broke the connection.

Annja couldn't believe it. No "goodbye." No "see you later." He just hung up. The man was insufferable. She truly didn't know why she bothered.

CASABLANCA MOON SAT at anchor out in the harbor. She was a dhow, an Arab-styled ship with a high keel and lateen sails. According to the information Annja got from the sailors she'd talked to, *Casablanca Moon* was an ocean-voyaging dhow.

Annja had to pay a water taxi to take her out to the boat. As she sat in the prow of the small motorboat, Annja surveyed the boat. Equipment maintenance, especially on a boat that was constantly exposed to the elements, spoke highly of the man she'd be hiring with it.

Five men sat in the stern in folding chairs. They drank beer from bottles.

"Ahoy the boat," Annja called up when the taxi reached the dhow. She settled her backpack across her shoulders and stood.

A man peered at her over the side. He was a few inches over six feet in height. A white, ribbed T-shirt that had almost turned gray encased his broad shoulders. Khaki cargo shorts wrapped his narrow hips. His skin was so black it glistened blue in the bright noonday sun. His head was shaved but he wore a small, neat goatee. Gold earrings glinted in both ears.

"Ahoy," the man replied in a deep voice.

"I'm looking for Captain Hakim Shafiq," Annja replied.

"What business do you have with him?"

"I want to hire this boat."

The man appraised her, then took a slow sip of his beer. "No offense, miss, but you don't look like you have enough money to hire this boat."

"I think I do." Annja met the man's direct gaze fully.

The man rubbed his chin. "Even if you do have enough money, what do you want with this boat?"

"I want to excavate a shipwreck."

The man grinned and looked back at the other sailors on the deck. "You got gold fever, miss?"

The other sailors laughed.

"No." Annja reached into her backpack and took out the small *naga* box. Casually, she tossed it up to the man. The cube spun and the gold twinkled in the light. "I think I've got an interesting find."

The man flicked a hand out and caught the cube. His hands closed over it and made it disappear.

"What do you think?" Annja asked.

The man extended a hand. "I'm Captain Shafiq. Welcome aboard *Casablanca Moon,* miss."

"SHIVAJI HAS a long association with piracy."

Fleet glanced at the files he had open in front of him.

They were back in Ranga's small, cramped office. Pictures of the man's wife, children and grandchildren covered the walls.

"Shivaji's only forty-seven," Fleet observed. "You make him sound ancient."

"I was referring to his family," Ranga explained. He lit an evil-smelling clove cigarette and waved the smoke away. "His mother's ancestors have been pirates since the eighteenth century."

"They sound committed."

"They are. They're descended from Kanhoji Angre."

Fleet shook his head. "Never heard of him."

"There's no reason you would have. Shivaji, as it turns out, is quite enamored of his nefarious ancestor. Kanhoji was fired by the British East India Company in the late 1690s and chose the path of entrepreneurship."

"By attacking his ex-employers' ships, I take it," Fleet said.

Ranga blew out smoke and grinned. "Kanhoji did know which ships carried the best cargo, you see. He was quite successful in his endeavor. He operated from a stronghold he called Severndroog on Vijayadurg Island."

Recalling the name, Fleet found a map in his files and located the island. It was south of Bombay. "It gave him a good location to strike from."

Ranga nodded. "Kanhoji had several other bases out there, as well. All of them were hidden places where he kept ships and cargo he seized. By the time he died in 1729, his pirate hoards numbered in the hundreds."

"Sounds like a very successful enterprise," Fleet commented. The old love of nautical stories swelled within him. His father had been a naval officer. His mother had often claimed that the sea flowed in their veins rather than blood.

"It was. Even after Kanhoji's death, the pirate empire lasted another twenty-six years under the leadership of his sons. The British didn't fully rout them until 1755."

Fleet leaned back in his chair and tried to get comfortable. There were days when he was more conscious of the prosthesis than others. Today he felt the old, gnawing absence of feeling anything beyond the end of his stump.

"I tell you this because we must be careful in our approach to Shivaji," Ranga said. "Not all of his ancestor's strongholds were identified. I suspect that Shivaji may yet have one or two hidden bases that we know nothing of. The sea, any sea, is a large and dangerous place. But the Indian Ocean is particularly so because of all the pirates operating here."

Fleet glared at Shivaji's picture. The man had only been arrested once, and the charges hadn't stuck. "We can't place that pistol in Shivaji's hands without your witness. Right now we don't have any leverage. How do you think he'd react if we leaned on him?"

Ranga released another puff of clove-scented smoke. "He would laugh at us." A small shrug lifted his shoulders and dropped them. "Perhaps there might be some anxiety. But Shivaji, if he is the one who has been using that pistol these past few years, isn't afraid of getting blood on his hands."

Silently, Fleet agreed.

"I do, however, know of another tactic we might employ," Ranga suggested.

"I'm all ears," Fleet responded.

"Good." Ranga smiled. "Perhaps because he has such a history of piracy and because that life is always so full of legend and superstition, Shivaji is prone to believing in tales of lost treasure."

Fleet considered that for a moment. "If you're thinking of salting the mine somewhere and trying to lure him to us, that's going to take a lot of time."

Ranga shook his head. "No. As a matter of fact there's an interesting situation that's developed all on its own. Have you heard of the American actress Annja Creed?"

15

Annja was in heaven. Naked, she stood in the opulent bathroom and stared at the huge whirlpool-equipped tub. The scented water started relaxing her at first breath.

She climbed into the tub and sank into the water. Heat just barely short of discomfort seeped into her. There was almost enough water to float. She enjoyed the weightless feeling until she couldn't hold her breath any longer. Slowly, she surfaced and felt the brisk chill of the room slap her.

For a time, she luxuriated in the water. Even though she relaxed her body, her mind stayed active. Just as she was forcing her mind to go blank, the telephone in the bathroom rang. She answered.

"I was told that you'd called," Lochata said.

"I did," Annja told the professor. "Are you still interested in helping me search the shallows for a shipwreck?"

Lochata sighed tiredly. "Over half of the class has

elected to abandon the dig. Given everything that they've been through, I don't blame them. It appears I'm left without a crew."

"I'm sorry to hear that," Annja said.

"The Shakti site is out here, Annja," the woman declared fiercely. "We found it. We have pictures of it. Once this matter is resolved and I have another class at my mercy, the university won't hesitate to allow me to try again."

"Good."

Motor noises sounded in the background.

"Where are you?" Annja asked.

"On a coast guard ship. They managed to send it for the rest of us. I'm told we'll be in Kanyakumari before nightfall."

"Do you have a room?" Annja asked.

"No. I plan on getting one—"

"Don't," Annja interrupted. "I have a suite. The couch pulls out. I'll take that and you can have the bedroom."

"I couldn't impose."

"It's not an imposition. I insist."

"Well, then, thank you," the professor said, sounding weary.

"You're welcome."

"Did you get the boat, then?"

"I did. Captain Shafiq and I struck a deal earlier. We made a list of things we'd need."

"At least we'll be able to rest while he gets the boat ready. I can't wait to bathe," Lochata said.

Annja smiled. "Call when you get here. I'll arrange for a taxi."

"You needn't do that."

"It's not me. It's the television show I work for. Since they're burying me, I'm blackmailing them into providing the room and the boat. Taxi fare is an incidental."

WHEN SHE CHECKED the archaeological lists, there were a number of emails from people who had seen the *naga* images. Several of them were simply requests for more information when and where Annja discovered it. Those were from university professors, amateur archaeologists and museum curators.

There were several that accused her of falsifying the images. Those were also from university professors, amateur archeologists and museum curators.

One person pointed out that until she could pinpoint what religion the *naga*s came from, she'd have a hard time figuring out what they might mean.

Annja had figured that much. India's history, culture, politics and religious beliefs were muddied from all the constant trade that had beat a path—or sailed the Indian Ocean—to their door.

She read on.

You're only dealing with a one-headed *naga*. That's gotta mean something. Many of the *naga*s over in that part of the world have multiple heads. Malaysian sailors think *naga*s are multiheaded dragons.

It's interesting that both of those objects have moveable parts. That's not usually the case. Unless they have some special purpose.

The ring's especially interesting. Maybe it was a

good-luck charm, but I get the impression there's another use for it. Someone obviously took good care of it. Good luck, and lemme know what you find out. I'm curious.

So am I, Annja thought. She sorted through the e-mail for another fifteen minutes before she found another interesting one.

Quite the cool find you've got there. Looks really old, too. I'm a linguist working on my doctorate, so I've seen a lot of languages. My favorite region is India.

Even though I tried for hours, I couldn't make heads or tails of that language. I even handed it over to the linguistics department here. So far none of them have any clues about the language, either.

That was odd, Annja thought. If the writer had experts studying the language and couldn't crack it, the objects were even older than the books.

She moved on to the next e-mail.

Don't know what you're looking into over there, but there's something you need to be aware of.

You posted those images on the Net, which is cool, but it could come back to bite you. Check out the story I've linked for you.

Be warned. You might want to keep a low profile over there.

Curious, Annja clicked on the link and opened the page in another window.

Snake Collector Murdered
Bombay, India.

William Newton, fifty-seven and noted American antiquarian and collector, was found murdered in his hotel room. Police investigators say that Newton was in the city to arrange the purchase of a *naga* statue for his collection.

The *naga* is a mythical creature that is half woman and half snake.

Newton's family told police that he had gone to Bombay to pick up the *naga* statue after he purchased it over the Internet. Police speculate that Newton was killed shortly after taking possession of the statue.

Police are still searching for the seller, a man described as in his mid-thirties, black hair and beard.

Anyone having any information is requested to contact police at once.

A quick check of the article's publishing history showed that it had originally run three years ago. Annja returned to the posting.

The thing is, this story isn't the only one. I've been able to find five more just like it. If you want, I can link those for you, too.

Thought it might be something you'd want to know.

Annja closed her computer and slid it away. She hadn't turned over the naga items she'd recovered. It wasn't exactly breaking the law. More like *delaying* it, she thought.

Besides, if she found the shipwreck, she might need them. Those items might be part of a collection. It would be better to have the whole thing.

16

Goraksh walked through his father's warehouse. He'd never thought of it as the family's warehouse. That would never be possible.

Crates were stacked throughout the small building. All of them held tourist trinkets that were shipped to various stores and shops along the coast where the cruise ships put in to port.

His father's business was modest and didn't draw much attention except occasionally from the local police. They launched periodic raids, but they never found anything.

A few years earlier one of his father's employees had made the mistake of bringing a small amount of hashish into the warehouse. When Goraksh had heard about it, he'd known the man was going to sell it to a friend. Goraksh had only been a handful of years younger than the man and had even been friendly with him. They'd swapped bootlegged American rock and roll.

When the police had come, Rajiv was forced to retain a barrister to keep the trouble from rolling over onto him. The police, once they discovered they couldn't make a case against Rajiv, released the young man.

Embarrassed but still in need of a job, the man returned to the warehouse. Goraksh hadn't known until later what his father had done. But the next morning the man was found out in the bay cut to pieces in a terrible "boating accident."

No one ever brought contraband materials to the warehouse again. They'd learned never to steal from Rajiv Shivaji.

None of the men in the warehouse spoke to Goraksh as he took the narrow wooden steps up to his father's second-story office. Only a few of them were pirates like Rajiv. Most only knew about the tourist-trinket business.

At his father's door, Goraksh took a deep breath and knocked.

"Come."

Goraksh twisted the knob and followed the door in.

Rajiv sat behind the desk. If he wasn't out on the antiques floor hand-selling curios and cheap authentic Indian furniture, he was in the warehouse helping move crates. When he was in the office, he was at the computers managing the business.

Goraksh knew no one worked harder than his father. Many people, some of them honest shopkeepers, praised Rajiv Shivaji for his work ethic.

The office was cramped and cluttered, but Goraksh knew his father knew where everything was. Books on appraisal, history, a dozen different pottery fields, more on gems and jewelry, filled the shelves. All of them had been well read. Papers filled with notes and drawings crowded the pages.

Goraksh was also convinced that his father, though he'd never finished public school, could have taught more history than most university professors.

Quietly, Goraksh stood and didn't take one of the two mismatched chairs in front of his father's desk.

Rajiv turned from the computer screen. "I have an assignment for you."

Goraksh nodded. He'd gotten the message on his cell phone after his last class.

"How is your neck?" Rajiv asked.

"It's fine, Father. Thank you for asking." In truth, if it wasn't for the painkillers Goraksh was taking, he didn't know if he could handle the pain.

"You are checking for infection?"

"Yes. There's none." Goraksh stood there awkwardly. He knew his father truly cared about him. That was what made the idea of separation or betrayal unfathomable.

"Good. It's better if we don't have a physician treat that. Too many questions would be asked." Rajiv reached into a drawer and pulled out a thick envelope. He riffled through it quickly, then tossed it to Goraksh.

Goraksh caught the envelope and peered inside. It was filled with rupees. Fear ignited within Goraksh. If this much money was involved, surely his father was going to ask nothing legal of him.

"What am I to do with this?" Goraksh asked in a thick voice.

"Four young men about your age were jailed today,"

Rajiv said. "I've been told they saw this American archaeologist, Annja Creed, with solid-gold *naga* statues from the sea after the tsunami struck."

Goraksh had heard no such thing, but he did know of Annja Creed. *Chasing History's Monsters* had several fans at the university.

"I want you to get these young men out of jail so that I may speak to them."

"Do you want me to bring them here?"

"No." Rajiv frowned. "Simply talk to them. Find out what they know about the *naga*s." He tapped the computer screen. "This Annja Creed is curious about them. *I* am curious about them."

Curious himself, Goraksh glanced at the computer screen. Images of *naga*s held steady there.

Rajiv had long been interested in the *naga*s. There were old stories he had discovered in Kanhoji's pirate treasures that spoke of an island of *naga* worshipers that had disappeared beneath the waves of the Indian Ocean thousands of years ago.

Personally, Goraksh believed they were just stories, not histories in any way. But he would never tell his father that.

"One thing further," Rajiv said.

Goraksh waited.

"Find out what you can about the Creed woman. I want to know where she's staying."

THE PART OF WORKING for the IMB that James Fleet most hated was the waiting for things to happen. When he worked with the special boat service they had to wait between missions, as well, but those times were filled with training.

He and his mates had spent grueling hours pushing their bodies to their physical limits. They'd run miles, exercised and fought in various martial-arts styles. He missed those times. His body still craved the physical release of pushing himself to the edge.

Instead, he sat in Ranga's office and went through the files he'd gotten familiar with over the past few months. The .357 Magnum tied together several of the cases he was currently working on, but he had other investigations on his plate. There was no shortage of work.

Idly, he reached down to massage the leg where the prosthesis was attached. In the beginning he'd been encouraged to massage the leg to ensure proper blood circulation. Now he thought it was just a habit.

Or a way of reminding himself that he wasn't what he had once been.

He opened the newest file in the assortment he had. A publicity still from *Chasing History's Monsters* lay on top. He'd decided immediately that Annja Creed was a good-looking woman. But researching her name had brought up a history of violent encounters.

That had set off Fleet's personal trouble detector.

Some of it could have been discounted as bad luck. Being in the wrong place at the wrong time had accounted for a lot of unpleasantness in people's lives. Annja Creed just seemed to be a magnet for it. There were domestic investigations concerning her, as well as international ones.

Some of it might have had to do with her work. As an archaeologist, she was always going off the beaten path and dealing with disreputable people. As a host for a television show that sensationalized the weird, exotic and murderous, she would attract even more.

But there was no accounting for the sheer volume of incidents the beautiful Ms. Creed had been involved in. More than that, in most cases he felt certain he was only getting half the story.

Ranga entered the room. The bags under the little policeman's eyes had become more pronounced.

"It appears," Ranga said, "that our patience is about to be rewarded."

Fleet closed the file. "Rajiv Shivaji came to talk to the men who attacked the archaeological team."

"Not Rajiv," Ranga corrected. "His son. Goraksh. But you can bet he is here at his father's behest."

Fleet smiled a little. There was nothing like watching a target take the bait.

GORAKSH FELT awkward as he sat in the jail's waiting room. Security cameras and the jailer on the other side of the bulletproof glass kept constant watch over him.

The chair was uncomfortable and his wound throbbed miserably. He forced himself to sit straighter but it didn't help.

Then the door opened and four young men stepped through looking rumpled and bruised.

Opening the papers his father had included in the packet of rupees, Goraksh matched the faces of the men to the ones on the paper. In the pictures, their bruises weren't quite so apparent.

The four men talked with the police officer who took off their handcuffs. The police officer pointed at Goraksh.

One of them walked over and scanned Goraksh from head to toe. "Who are you?" he demanded.

"I'm the man who's taking you from this place," Goraksh answered. He'd grown up around harder men than these. His father had taught him how to deal with them.

"Why would you do that?"

"I know someone who wants to employ you."

The four men looked at each other, then back at Goraksh. "Who?"

He named the amount his father had first told him to offer them. There was negotiation room, of course. There always was. But the men accepted straightaway.

"What are we supposed to do for that?" one of them asked.

"Just answer a few questions. If there's anything further, you'll be compensated for that, as well."

On the other side of the glass, Goraksh watched a white man in plainclothes step into view. He looked American or European. Or perhaps he was Russian. He was too fair-skinned to be anything else.

For a moment his gaze met Goraksh's and held it. Goraksh was afraid. Then the man turned his attention to one of the computer screens.

Breaking the thrall that held him, Goraksh led the way out of the building. The four men behind him hesitated for just a moment as they considered their options.

Goraksh knew he couldn't return to his father empty-handed. "Gentlemen," he said patiently in a tone that he normally used at university.

They looked at him.

"You are in a lot of trouble," Goraksh pointed out. "Perhaps it would be a wise thing to have some money in your pocket before you tried to run. This will only take a few minutes."

Reluctantly, they agreed.

Goraksh led the way to the van that he had parked nearby.

BY NINE-THIRTY THAT EVENING, after dinner with Professor Rai, Annja had set aside her computer and tried to turn her attention to television.

On the way back from the restaurant downtown that Lochata had recommended, Annja had spotted an entertainment store advertising the first season of *House*. She'd bought it on impulse and discovered to her delight that Lochata loved American television but usually didn't have time for it.

Since both of them had napped, Annja after her bath and Lochata aboard the coast guard ship, neither of them was ready for bed.

They were on the second episode of the first disc when a loud knock sounded on the door.

Annja paused the DVD player and peered through the peephole. When she saw Captain Hakim Shafiq standing there scowling, she knew there was a problem.

17

Annja opened the door. "Captain."

"Miss Creed." Shafiq looked angry and his deep voice carried out that impression. "We need to talk." He barreled past her and stepped into the hotel suite.

Annja curbed the impulse to grab hold of the man and throw him back out into the hall.

"Won't you come in?" she said sarcastically to his back. She peered out into the hall to see if there were any other surprises, but nothing appeared in the offing. After she pulled back into the room, she locked the door and turned to face her uninvited guest.

Shafiq looked confused. "Didn't know you had company," he said.

"I do."

"I'm sorry about barging in like that." Shafiq nodded at the professor. "Ma'am."

Lochata nodded. "Clearly you've come on a matter of some importance."

"I have. Very important."

"Can I get you something, Captain?" Annja asked. "Water? Tea? Maybe a refresher course on civility and politeness?"

Shafiq grimaced. "You've put me in a most perilous position, Miss Creed. Especially by having to get my boat ready on such short notice."

"How have I done that?" Annja felt less defensive and more curious.

"You gave me purchase order numbers to work off to get the necessary supplies."

Annja nodded. She'd arranged that through Doug Morrell.

"They've all been canceled," Shafiq said. "Every last one of them. Suppliers have been calling me all evening long. All those things that are supposed to be delivered tomorrow? Well, they're history, most of them. And the rest I have to pay for out of my own pocket." He glared at her. "Now, I don't know what kind of game you have running, but you're not going to make your problem my problem."

"DOUG," ANNJA SAID as evenly as she could under the circumstances, "I know you screen your calls. I know you're probably sitting there listening to this one. We need to talk."

The white noise of the recording session continued for a moment, then the call ended.

Annja stood in front of the window overlooking Kanyakumari and the harbor. Lights filled the night. Anxiety rattled through her.

I do *not* want to lose this opportunity, she told herself. I'm not going to lose it.

She punched Redial. "Doug, if I don't hear back from you in the next five minutes, I'm going to bury myself in archaeology work and you're going to have to find a new cohost for *Chasing History's Monsters*."

It wasn't an idle threat. The exposure she'd received on the television show had spawned a lot of offers from other places. But a new endeavor, where she didn't already have contractual agreements, could have eaten up her time and taken her completely from the real work she wanted to do.

She broke the connection and looked back at Shafiq and Lochata. "He'll call," she told them confidently. "I've never threatened to quit before."

Shafiq took a sip of water from the bottle she'd given him. He didn't look impressed.

"Perhaps I could arrange something at the university," Lochata said. "Of course, it wouldn't be permanent, but I've learned when these things happen it's better to have something to shore you up."

"I'm not fired," Annja said. "They wouldn't fire me." But deep down she didn't know if that was true.

The phone rang.

"Doug, it's about time. If you think for one minute that I'm going to put up with—"

"It's not Doug," Roux interrupted. "What has you in such a dither?"

Annja folded her arms and resumed pacing. She

couldn't believe everything was just within her reach and it was going to be yanked away.

"I'm dealing with a stressful situation at the moment," Annja said.

"Oh. Well, it doesn't sound like you're dealing very well with it."

"I'm dealing with it just fine. What do you want?"

"A more civil conversation, perhaps," Roux said.

Annja sighed. "Sorry. Caught up in the moment. Maybe it would be better if we talked at another time."

Roux ignored the suggestion. "Have you found out anything more about the *naga* figures you found?"

Some of Annja's irritation turned Roux's way. He had to be the most selfish man she knew. Everything in his world turned around him, and he expected everyone else's world to be accommodating.

"Nothing," Annja said.

"What have you been doing with yourself?" If Roux had noticed her irritation, he plainly didn't care. More than that, he sounded irritated himself.

"I've been researching. I just haven't turned anything up."

"Well, you should pursue this," he said.

Annja gazed out at the harbor. She thought she could see the *Casablanca Moon* out in the distance, but the night was so dark she couldn't be sure.

"I'm trying," she replied. "That's why I called you earlier. Are you interested now?"

"Slightly intrigued."

"Why?" Annja wondered what had changed his mind.

"Have you been able to translate the language at the base of the figure?" Roux asked.

"No."

"Neither can I."

"That intrigues you?"

"Yes," Roux growled.

"You're not a linguist," Annja said.

"No, but I know people who are."

That stopped Annja. It was true. Roux knew people everywhere.

"None of those people can translate the language, either," Roux went on. "I'm told that the Indian languages derive from the Dravidian and Indo-Aryan."

"They do." Annja recalled that from her studies, though that wasn't within her field of expertise.

"It's been suggested that this is a language *isolate*."

"A natural language?" Annja asked.

"Exactly. I wasn't familiar with the term," Roux said.

"A language isolate usually consists of a family of one. No other languages touch on it. The Basque language in Spain is an example."

"That language is spoken by millions of people," Roux pointed out. "The possibility exists that this is a dead—or at least a forgotten—language."

"Wow," Annja said.

"You say wow. I say infuriating," Roux replied.

"But this could be really cool."

"Clearly."

Annja thought quickly. Roux had a lot of money. She needed money to make the excavation happen.

"How intrigued are you?" she asked hopefully.

"Not so much."

"How can you *not* be interested?"

"This isn't my thing, Annja. There's no great mystery here."

Annja rethought her position. If I can't solicit him, there's always begging.

"I have a problem," she said.

Roux sighed. "If this is going to be one of those soul-baring moments, I'm really not interested."

Why? Annja wanted to ask. Is the masseuse waiting?

"I need money," Annja said.

"You have money," Roux replied.

"Not enough." Annja thought of Roux's mansion in Paris, the private airplanes and the wealth she'd seen him throw around when he had to. However, he was extremely frugal by nature.

"What do you need money for?"

"Boat rental, supplies. For the excavation."

"How soon would you need it?"

Hope sprang within Annja. "Tomorrow morning?" She looked at Shafiq.

The big captain hesitated, then nodded.

"Tomorrow morning is impossible. You're talking about a large amount of money."

Annja knew that was true. "I want to do this excavation, Roux."

"Again, I don't see how that concerns me."

"What if I had a life-threatening condition that required an operation?"

"Then," Roux said cautiously, "we would talk. But you don't have a life-threatening condition requiring an operation. And if you did, there are always payment plans."

Annja wanted to kick the glass out of the window and scream at the dark heavens. How could he be so infuriating?

"There's no way you're going to extort money from me, Annja," Roux said.

"Consider it a loan. I'll pay you back." Surely she could package the story rights in some way, especially if she was no longer involved with *Chasing History's Monsters*. That would pay back whatever financial obligation she incurred.

"You could be dead tomorrow."

Roux's flat pronouncement caught Annja off guard. A chill raced up her spine.

"That's something I have to remember at all times in my dealings with you," Roux said in a softer voice.

And what do you say to that? Annja wondered.

"The main reason that I say no," Roux continued, "is that moving around that kind of money that quickly could leave me in a precarious position. I've been very careful with my assets. I don't want to draw the attention of international money mongers."

"Sure," Annja said. "I understand." But she didn't. She'd never had that kind of money to throw around.

"I would be interested in hearing the end of this tale," Roux said. "Perhaps, when you're finished over there, we could get together over dinner. I've some things I'd like to discuss with you."

"Right." Annja's eyes burned and she knew it was from all the disappointment she was feeling. What do you think? she asked herself. That Roux's going to ride in to save you like a real father every time you get into trouble? He didn't even do that for Joan.

"I wish you the best in your endeavor," Roux said.

Annja thanked him and hung up.

"No go, eh?" Shafiq asked.

Not trusting her voice at the moment, Annja shook her head.

"Well, then, I guess we're done before we got started. We both have our problems, Miss Creed." Shafiq stood. "I shouldn't have let myself get swayed by all your pretty words. I've been in this business long enough to know to go strictly cash on the barrelhead."

Annja took a deep breath and let it out. "Wait. There's one more call I can make." She punched the buttons on the phone, hoping she recalled them correctly.

The phone rang three times, then Garin answered. "Ah, you memorized the number from the other phone."

Annja knew she was in trouble from the smug tone in Garin's voice.

18

"I was told you were going to throw that phone away," Annja said. The thought crossed her mind that he hadn't thrown it away because he'd guessed she might call again.

"You called just to test that?" Garin's voice held a mocking tone.

"No." Annja was too conscious of Shafiq's and Lochata's attention on her. When she glanced at them, they both looked away.

This is not a personal call, she wanted to tell them. This is strictly business. They had detected something in her words or body language. But it was something that definitely wasn't there. The situation was simply embarrassing. She turned back to the window and tried to ignore them, but then she caught herself glancing at their reflections in the dark glass.

"Then why did you call?" Garin asked.

"I want to present a business venture."

"Okay, I'm listening."

"I think I've found a shipwreck."

"You believe this is where you got the little trinkets involving the *naga*s?"

Wariness consumed Annja. She didn't know who was trying to trap whom now. Garin had been along on other expeditions she'd done. He had a habit of picking up treasure and absconding with it when the chance presented itself. Losing what might be out there wouldn't be a good thing. The Indian government authorities would frown on such losses occurring under their noses.

Garin also took the occasional opportunity to try to kill Annja and Roux. The rescue earlier in the day had been the most surprising thing he'd ever done. Except for the time he'd saved Roux's life instead of letting him die.

"Yes," she said. "I do."

"I saw them online," Garin said. "I know your usual Internet haunts, and from the preoccupied manner in which you dealt with me—"

"Don't forget that I'd just been told I'd been declared dead."

"—I knew you were working on something."

Annja could hear the smug superiority oozing through Garin's words.

"I am working on something," she admitted.

"The *naga*s."

"Yes."

"The business venture you're offering has something to do with the *naga*s?"

"And the shipwreck I believe is out there."

"If you're going to proposition me, let's hear it."

Annja didn't much care for his choice of words. Garin knew enough languages that she was certain the choice had been deliberate.

"I want to attempt to recover whatever's left of that ship," she said.

"Why?"

"Because it's what archaeologists do."

"I'm not an archaeologist. And if the shipwreck is found in territorial waters, which is what I'm guessing, then salvage can't be recovered without cutting the Indian government in. I've no intention of doing that."

"The Archaeological Survey of India is going to get whatever we find," Annja said. "Eventually," she added.

"Then there's no profit," Garin stated.

"Don't you own a media company?"

Garin paused. "Someone's been snooping around my companies."

"I wouldn't call it snooping," Annja said.

"Then what would you call it?"

"Curious. You'd be surprised what you can pull up on Wikipedia."

Garin laughed explosively. When he finally regained control of himself, he said, "I do like you, Annja. I swear that you're one of the most outrageous women I've ever met. And I've been around a lot of women."

Annja felt good about that. She didn't want to, but she did. She didn't want to measure up on any of Garin's scales.

"You own a media company," Annja said again.

"I do. In France."

"I propose to let you have the film rights to the recovery of the shipwreck excavation," she said.

"With you in the lead?"

Annja hesitated. She'd have to talk to her lawyer to make sure she wasn't breaking any agreement with *Chasing History's Monsters.* But she believed she was on safe ground. She knew she could take on film work as long as it didn't directly compete with anything she was doing with the television show. And the television show pulled out of this one, Annja reasoned. No conflict of interest.

"Yes," she said. "With me in the lead."

"You'll handle the excavation from start to finish?"

"I will."

Garin was silent for a moment. "I don't know."

"It's a good deal," Annja said. "I'm a proven commodity in the entertainment market. You can get a deal with HBO or Showtime, or The Discovery Channel or The Learning Channel."

"Probably. But I don't know if it's going to sell any better than *The Best of Annja Creed Collection.*" Garin paused. "And why isn't *Chasing History's Monsters* all over this?"

Annja decided to lay her cards on the table. She had nothing to lose. If she tried to hide the show's initial involvement Garin might be upset.

"They were," Annja said. "They pulled out their funding a few minutes ago."

"Why?" Garin asked.

"I don't know. They're not returning my calls."

"I see."

Annja listened to the silence for a time. Her anxiety was growing.

"I'll agree to your proposition," Garin said finally. "I'll provide funding—"

"Immediately," Annja interjected.

"Immediate funding," Garin amended, "for your project. We'll work out the points on the royalties later, but I assure you I'm going to get most of them."

"Agreed." Annja wasn't interested in the profits. She just wanted the opportunity to see what was there. If she'd been independently wealthy, she would have funded the excavation herself.

"There is one further condition that has to be met," Garin said. "Otherwise there's no deal."

Irritated, Annja asked, "What?" She was sure she was about to have the carpet yanked out from under her again.

"You have to agree to have dinner with me," Garin said.

"What?"

"Dinner," Garin said. "You. Me."

"We've had meals together before."

"This will be different," Garin said.

Welcome to my parlor, said the spider to the fly, Annja told herself.

"It occurs to me that if we had dinner, you might find it easy to kill me," Annja said. She saw Shafiq's and Lochata's eyes widen in their reflections in the glass.

This has to be one of the most confusing conversations they've ever eavesdropped on, Annja thought.

Garin chuckled. "You don't trust me?"

"I can't believe you have to ask," Annja said.

"You can pick the place," Garin declared magnanimously.

"All right."

"But I get to pick the city."

"That doesn't sound fair," Annja said.

"It's a deal breaker," Garin warned.

Annja thought quickly. She didn't trust Garin. She had plenty of reason not to. He'd tried to kill her and Roux on more than one occasion. Then, like this morning, he'd turned around and done something that benefited her.

Don't make mistakes, she warned herself. Garin never does anything that doesn't reward him.

"Well?" Garin asked.

Annja thought about the *naga* figures and the fact that an undiscovered shipwreck had probably been brought up from the ocean's floor after hundreds of years. She wouldn't let the opportunity pass her by.

"Deal," she said.

"Let me know the name of the man I need to talk to and how to reach him, and I'll take care of the financial matters," Garin said.

"As it happens," Annja said, "he's here with me in my hotel suite."

"Oh?" Garin's voice lost some of the devil-may-care resonance.

Annja took pleasure in that as she handed the phone over to Shafiq. The captain wandered off and started talking numbers with Garin.

"Everything is all right now?" Lochata asked.

"I think so. We've got a new financial backer."

"Ah. Is this an old acquaintance of yours?"

"Not in the way you think," Annja assured her.

"Someone you're interested in, then?"

"No, definitely not," she said.

"Of course not," Lochata said, but she looked as if she didn't believe Annja for a minute.

Annja told herself she wasn't interested in Garin, but she knew she was lying. He was always intriguing. Not only that, but what was Roux going to do when he found out?

19

"Look, Annja," Doug went on, "it wasn't my decision to pull the funding for the excavation. I was fighting for you. I'm *still* fighting for you. I'm writing a memorandum right now that's fighting for you hard-core."

"Don't strain yourself," Annja said. She sat at a small table in the cybercafé only a few blocks from the pier and listened to Doug Morrell plead his case. She almost felt sorry for him because he sounded sincere, but she was distracted by the excitement about getting the excavation under way.

Captain Shafiq and his crew were finishing up with the supplies and equipment. Lochata was stating their case to the ASI and pulling in favors from the university. They believed they'd be ready to set sail by 2:00 p.m. That was less than three hours away.

"Doug," Annja said, "it's no big deal. Don't sweat it."

"Don't sweat it? After you strong-armed me to get the money for your spelunking trip?"

"Marine archaeology," Annja corrected, "is more like diving. Spelunking has to do with caves. We're not in a cave." She took a bite of her sandwich. "Unless there's a cave down there on the ocean floor. Then it's diving and spelunking. But as of right now, there's no cave."

Doug took a deep breath. "You don't want to leave the show."

"I'm tired of chasing monsters, Doug," Annja said. "It gets in the way of archaeology."

"Not always."

"Most of the time."

"Everywhere we send you, you find time to archae-ologize," Doug said.

"That's not a word." Annja felt good about having the upper hand for a change.

"Whatever. The point is, I know we've sent you to places that you would never have been able to go. And while you were there you've always found really cool stuff. Admit it. You know it's true," Doug said.

"Are you trying to play hardball with me?" Annja asked.

Doug hesitated. "Yes. Is it working?"

"No. I just wanted to make sure. I don't hear you do that often. It's kind of cute."

"Cute?" Doug spluttered in indignation.

"Endearing?" Annja offered.

"We *need* you on the show," Doug said. "It isn't just Kristie's, um, *assets* that sell *Chasing History's Monsters*."

Now, that was a pleasant surprise. Everybody she talked to felt that Kristie Chatham and her *assets* were the major attractions of the show.

"Put that in writing," Annja said.

"What?"

"What you just said. The asset thing. Put it in writing."

"You're kidding."

"Am I? Television stars tend to make outrageous demands. I'm practicing."

"You've got to be kidding. I can't do that."

"But you can kill me and put out a memorial DVD."

"Hey," Doug said quickly, "I wasn't the one who killed you."

"The memorial DVD was your idea," Annja said.

"Okay, before we get off track here, the whole problem was that memorial DVD. Once everybody found out you were still alive, the funding immediately got cut."

Annja was surprised. She hadn't told anyone she was still alive who didn't already know. "I told you keeping something like that secret was a mistake," she said.

"Yeah," Doug said bitterly, "but no one expected you to be the one that blew the whistle."

"What are you talking about? No whistle-blowing here," Annja said.

"I'm not saying you did it, but it sure looks like you did it."

"What looks like I did it?"

"The Internet footage."

"What Internet footage?" Annja asked.

Doug gave her the address and she quickly entered

it in her browser. Her computer only took a few seconds to cycle through and bring up the link.

A window opened in the center of the screen. Words scrolled across as a clock ticked down.

Annja Creed, star of *Chasing History's Monsters* dealing martial-arts mayhem! See her deadly kung fu grip!

The camera opened up on Jason in a hotel room. He was all smiles.

"Hi. My name is Jason Kim. I'm an archaeology student from UCLA." He grinned again. "I'm single but looking. I like music and long, slow kisses. You can find my e-mail here." He pointed and his e-mail address suddenly appeared in front of him: onehotchickmagnet@yourdreams.org.

Annja rolled her eyes.

"Anyway, I'm presently in India on a dig for some really cool sacrificial victims," Jason said. "I got to dig up some of the bodies. Way cool. Our camp got destroyed by a tsunami. Not so cool. A few people would have died, but Annja Creed, the archaeologist from *Chasing History's Monsters,* saved them. Watching her do that was cool squared."

The scene changed and showed the speed boat the four would-be robbers had arrived in.

"The next day, these guys came up and wanted to rob us. What you're about to see is amazing. Annja Creed comes up out of the water—"

A yellow line encircled Annja as she vaulted from the water.

"—and proceeds to kick these guys' butts!"

On the screen, Annja defeated all four men again in rapid succession. The image froze on her as she pointed the pistol at the man's head.

"Her studio is telling everybody she's dead," Jason continued. "They're even selling a bogus memorial DVD. Don't fall for that. She's still alive and is planning on trying to find a shipwreck she thinks these came from."

The view cut to the *naga* statues.

"I'm going on the marine excavation," Jason said as the camera panned back to him. "Professor Rai was looking for volunteers, so I volunteered. Hopefully I'll have more news soon."

Annja leaned back in the seat. "I don't see how this little bit of—"

"CNN, Entertainment Tonight and Fox News all got hold of it," Doug said.

"Oh."

"Naturally, all those memorial DVDs we've been pressing aren't going anywhere. The people who ordered them want their money back."

"I can understand that," Annja said.

"Marketing wants my head. Nobody wants to refund those orders. That's gonna cost even more money just handling all the paperwork. We're gonna blow budget really big for a few months."

Annja conceded that was a problem. That meant *Chasing History's Monsters* wouldn't be sending her anywhere exotic for a while.

"Nice bikini, though," Doug said.

Annja smiled. "Thanks. It's new."

"Very stylish." Doug drew in a deep breath and let it out. "So you can see the position I'm in. The memorial DVD is dead in the water. If I have to go tell corporate you're breaking your contract with us— Well, I don't want to do that."

"Then don't," Annja said.

"Huh? What happened to the whole strong-arm approach?"

"I don't need it. Someone else is going to finance the dig." Annja left money for her meal, put her computer away, slipped her backpack over one shoulder and headed out into the street.

"That's great," Doug said.

It's not going to be great when you realize I signed away rights to the dig, Annja thought. But she didn't want to deal with that at the moment.

Annja walked toward the docks. Sudden movement in the alley to her left drew her attention. When she glanced in that direction, she saw some familiar faces.

All four of the men from the speedboat were in the alley. They'd brought four friends.

"Doug," Annja said calmly, "I'm going to have to call you back."

20

Annja's first impulse was to run. Her martial-arts training supported that course of action. When confronted with adversity, she'd been taught to seek out the path of least resistance.

She shoved the cell phone into the backpack pocket and hitched the straps over her shoulders. She dashed along the street.

A small sedan rocketed across the street, narrowly missed a woman with two small children and shrieked to a stop in front of Annja. She stopped only a few feet from the car and looked around.

Three armed men with pistols got out of the car.

Annja didn't know if the men would actually shoot at her, but she couldn't take the chance. There were too many citizens and tourists on the street including a lot of children.

"Hey, lady."

Hands on the straps of her backpack, Annja turned to face the young man who had called out to her. His face still showed bruises from their earlier encounter.

"The way I see it," the young man said, "you are not leaving this place unless I give you permission."

"I see you brought friends," Annja said. "That's surprising."

"What do you mean?"

"I didn't know you had any friends," Annja said.

The young man grinned, but the effort wasn't as menacing as he'd intended because his lips were still swollen. "You think you are funny."

No, Annja thought, I was just trying to stall for time and hoping that some good Samaritan noticed my dilemma and called for the police.

The man held out his hand. "Give me the backpack."

Annja looked for an escape route but found she was ringed by the men.

"Hurry," the man ordered. "Give me the backpack and I'll let you live."

"If you're looking for those *naga* statues," Annja said, certain she knew what the men were after, "I don't have them anymore."

"You're lying."

The man was right. She hadn't yet handed the things over to the ASI. Professor Rai had arranged for an indefinite loan of the items while Annja completed her research.

"You're not getting the backpack," Annja said.

"That's okay. I'm happy to take it from you." The man reached into his pocket and pulled out a long-bladed knife. He stepped toward her.

Annja stepped forward, seized the man's right hand in her left to block any attempt he might make to break free and unleashed a right kick that had plenty of hip in it. The man raised his free hand to block the kick, but it had little effect.

Annja's foot drove the man's hand into his face and knocked him to the ground. She held on to his hand long enough to strip the knife from it. Without hesitation, she tossed the knife out into the street.

The men swarmed in at her like a pack of jackals, but in their eagerness to get to her they got in each other's way.

Annja reached into the otherwhere and gripped the sword's hilt. When she pulled, it appeared in her hands. The sword was three feet of naked, unadorned steel. It was a warrior's instrument.

The sudden appearance of the sword made the men stop short. Their eyes widened as they realized their prey had gone from helpless to lioness.

Strength and confidence poured into Annja as she took a two-handed grip on the sword. She didn't want to kill anyone if she could help it. The police would frown on that and potentially delay the excavation. But more than that, she disliked killing unless there was nothing else she could do.

The young men were thugs but they weren't hardened killers. Annja had been in the company of both enough to know the difference.

She stepped forward again and drove the sword's hilt into the forehead of the man ahead of her. The man staggered. Annja whirled and delivered a back kick to his chest that hurled the unconscious man into the two men standing behind him. All three went down in a tangle of arms and legs.

One of the men had a machete and brandished it with enthusiastic menace. Annja easily knocked the machete from his hands with her sword. Her next slash sliced through the loose folds of the man's shirt. He looked down in shock, as if looking for his intestines to come tumbling out. He dropped to his knees and hugged the ground.

Annja whirled again and kept track of her opponents the way she'd trained to do in the dojo. She caught another man's temple with the flat of her blade hard enough to leave him disoriented. Still on the move, she rammed her left elbow back twice and hammered another man in the face. His nose broke and he fell backward.

The forward momentum of her attackers was shattered. One of them lay prone on the ground. Another was still trying to get his breath back and a third couldn't stand up straight.

The remaining men chose that moment to run.

Annja stood her ground and didn't try to pursue. She was relieved the guns had been for show and that no one had started a wild shooting spree. As she glanced around, she realized the attack had attracted a crowd. She turned and headed for the docks. When she released the sword, it fell from her hand and faded into otherwhere.

But the question remained as to who had sent the men. If they'd simply wanted revenge, they'd have taken the backpack as a matter of course after they'd beaten or killed her. They wouldn't have asked for it first.

She lengthened her stride and moved into a steady jog as she dodged traffic.

"AHOY, MISS CREED." Captain Shafiq stood on the deck of his boat and waved. His skin shone in the heat and perspiration dappled his gray shirt.

"Captain." Annja accepted the man's hand as she stepped onto the *Casablanca Moon*. "Any problems getting the supplies?"

"No, ma'am." Shafiq shook his head. "Were you expecting a problem? Your friend's money arrived in good order this morning."

"No problem there," Annja said. "But there may be another situation." She quickly told him about the attack.

Shafiq frowned. "You say you don't know these men?"

"Only that they tried to rob the excavation site."

The big captain surveyed the docks with a wary eye. "Maybe they were after the figurines you recovered."

"Perhaps. If that's true, this may not be the only time they try."

Shafiq nodded. "That would be their mistake. The men I have aboard this boat have fought for cargo before. If you make your living as a merchantman in these waters, you can bet you'll have to deal with pirates at some point. We have." He took a deep breath and let it out. "Most of them prey on the weak. We're not weak." He glanced at her. "Do you have reservations about this voyage?"

"I do," Annja admitted.

"Are you thinking of calling it off?"

"I'm not," Annja replied. "But I wanted you to know what you were dealing with."

"We can take care of ourselves, Miss Creed." Shafiq

touched the brim of his hat. "Now, if you'll excuse me, I've got to get this boat squared away."

AFTER SHE STOWED her gear in the small berth where she was staying with Professor Rai, Annja returned to the deck. She had limited experience aboard boats, but she had enough to help with storing provisions and some of the equipment.

Shafiq protested her efforts, but Annja pointed out that it was too hot to stay belowdecks and she didn't think wandering around town was a good idea. Reluctantly, Shafiq agreed and assigned her to assist one of the more experienced hands.

While on the deck, she could also watch the docks for any sign of the young men. She felt good about the work. The heat baked into her and warmed her muscles.

Lochata arrived only a short time later carrying several maps and books. Annja and the deckhand helped Lochata load her cargo onto the boat.

"Nautical charts and books regarding the area," Lochata said. "Perhaps we can use them to narrow down the search for your ship."

"It's not my ship," Annja said.

"If we locate it, the ship will be named as your find."

"You've cleared up all the paperwork with the ASI?" Shafiq asked.

Lochata nodded. "I had to call in some favors to push the paperwork through so quickly." She rummaged in the large document bag she carried and pulled out a sheaf of papers. "You'll find everything you need there that will answer any questions the coast guard or navy might have."

Annja took the papers and thumbed through them. "Good. I wouldn't want them to arrest me as a thief."

"They won't. Not with those papers. After I explained the circumstances of the tsunami and pointed out the fact that the sea could just as easily take back what she'd given—if she hasn't all ready—they capitulated." Lochata smiled. "It helps that they know they can trust me from previous dealings, and that you came forward with what you found. I don't think that it hurt that your presence will guarantee television exposure of the event."

LESS THAN AN HOUR LATER, Annja stood in the stern as Shafiq gave orders to the helmsman to take the boat out into the harbor. She shaded her eyes as she raked the docks for anyone who might be overly interested in their departure.

Now that they were under way, her perspiration-sodden clothing felt cool. She lifted the hair from the back of her neck and held it in one hand for a time to luxuriate in the breeze.

Other than a few friendly waves and wishes for safety and luck, the *Casablanca Moon* drew only cursory attention. But Annja didn't doubt for a moment that someone was out there watching. She could feel their gaze on her.

STANDING ON THE VIBRATING dock in the shadow of a load of cargo stacked on pallets, James Fleet watched the *Casablanca Moon* move through the harbor. He held his binoculars at an angle so the sun wouldn't catch the lenses. The dhow cut the water smoothly despite the equipment lashed to her stern decks.

He focused on the woman standing there looking back. She was very beautiful. He'd seen her a few times on television, but that kind of programming had never really appealed to him. He'd preferred military history. Occasionally, he'd watched *Chasing History's Monsters* when Annja Creed was focusing on the history and not some monster of the week.

Fleet broke his gaze. Some people could feel when someone stared at them too hard or too long. Fleet had always been able to. Even the day he and his team had been ambushed, he'd known he'd drawn the attention of someone dangerous.

Besides, he wasn't looking for the woman. She was just the bait in the trap.

After taking up a new position, Fleet turned his attention back to the docks. In less than a minute, he spotted Rajiv Shivaji standing in the prow of one of the small ships he owned.

Only a moment after Fleet had focused on the man, Rajiv shifted uncomfortably and turned to look back along the docks, as well. Fleet turned away.

Okay, you bloody sod, Fleet thought. You're interested. What are you going to do about it?

21

"A ship doesn't simply sink beneath the ocean when it goes down. A large part of what happens to a vessel beneath the waves depends on what caused it to sink."

Annja sat in the small ship's galley and listened to Lochata speak in the calm, measured voice that she doubtless used in the classroom. Her tone was almost soothing, but Annja concentrated intently on what was being said.

Captain Shafiq and the four men who'd been hired to assist in the diving crowded the table, as well. All of the local divers were young and fit. Two were Indian, one was Nigerian and the fourth was German. They all spoke English and had worked dives before. Some of them, Shafiq had acknowledged, weren't necessarily for archaeological purposes. But he'd promised they were all trustworthy.

"Are you familiar with Muckelroy's model?" Lochata asked. Her gaze lingered on the four young divers.

Annja knew about the marine archaeologist's template for site recovery, but she also knew Lochata hadn't been asking her. She sipped water and waited.

"Sure," one of the Indian men said. "He wrote out a study of what happens to a ship that's gone down and what you can expect from a salvage job. What happens to the ship itself, the materials it's made out of and what to expect if the shipwreck has been disturbed in the past by other divers or natural phenomena."

Lochata smiled in obvious pleasure. "Very good. And that's exactly what I want you to remember while we're down there."

The professor was also a certified diver and had stated that she would be diving, as well. After seeing Lochata deal with the hardships of the Shakti excavation, not to mention the tsunami, Annja didn't harbor any fears about the woman. Lochata was quite capable of taking care of herself.

"Before we go through that shipwreck, provided it's out there and we find it," Lochata went on, "we're going to film it and document its present conditions."

"Won't the tsunami have disrupted it?" the Nigerian asked.

"Without a doubt," the professor answered. "We're not going to get the best look we could have at this site, but we're going to make sure we do the best we can."

"This is not a treasure hunt," Shafiq interrupted.

"I heard there was gold down there," one of the Indian men said. "I was told she—" he nodded at Annja then "—found some gold figures and some coins."

He's well-informed, Annja thought warily. Of

course, information like that could be expected to fly through a port.

"Gold has been found," Shafiq told them. "And I'll have you remember that while we're working that site, we're working under the umbrella of the ASI. Professor Rai is their representative in this matter." His eyes turned hard. "If anything goes missing while we're out there, I'll see you jailed. You have my word on that."

"No prob," the man said. "But people are talking about this. We might not be the only people working that area."

"We will be," Shafiq said. "All I have to do is make a radio call to the navy and we'll have reinforcements in place in short order. In the meantime, I've hired men aboard this boat who know how to handle weapons."

"I was just saying," the man said defensively.

"We go slowly," Lochata said. "That's how we best learn what history has to teach us."

The man nodded.

"Muckelroy's model tells us a lot of what we can expect as we sift through the debris," Lochata said. "Of course the tsunami will have altered some of that. The first thing we want to discover is why the ship went down. Many ships aren't wrecked. They were abandoned and scuttled because they'd outlived their good years of service."

"That doesn't seem likely in this case," the German said. "They wouldn't send a boat to the bottom if they knew it had treasure aboard."

"The people who sank it might not have known," Annja said. "Captains and quartermasters often had hiding places aboard their vessels."

"In the same way the pirates and smugglers do now," Shafiq said.

"Yes," Lochata said. "How the ship went down will tell us a lot of what we need to know."

"Do you even know what kind of ship you're searching for?" the Indian asked.

"No," Lochata answered. "I hope that's one of the first things we discover. If it wasn't scuttled, the ship could have been beached somewhere then slid off into the shallows a long way from where it started out. But there have always been merchant wars and pirates in this area. I think we're more likely—since gold is probably aboard, as well as the skeletons that were recovered in the sail canvas—to find that the ship was sunk either in battle or as the result of a storm."

"If that's the case," the Indian said, "there could be more than one ship down there."

Lochata nodded. "That's what we want to find out."

DESPITE HER FATIGUE, Annja could barely sleep that night. Since he had the necessary equipment to permit night sailing, Shafiq continued sailing.

As she lay in the hammock, swaying gently with the boat's motion across the waves, she knew her mind was too busy to allow sleep. With a sigh, she gave up.

COOL AIR SWEPT in from the sea and caused Annja to wish she'd brought a jacket. She stood in the stern and gazed up at the full moon. Dark clouds scudded across the bright lunar face.

"Miss Creed."

Startled, Annja turned toward the voice and found Shafiq seated in a camp chair at the stern railing. The orange coal of his cigar glowed briefly and lifted his ebony features out of the shadows.

"Captain."

"What are you doing up?"

"I'm restless. I got a good night's sleep last night," Annja said.

"You also worked hard today."

Annja shook her head. "I guess I'm excited."

"I suppose that's allowed," the captain said.

Annja shivered.

Shafiq reached into a duffel bag at his side. "Here." He tossed her a wool jacket. "It's clean."

"Thank you." Gratefully, Annja pulled the jacket on and felt instantly warmer.

"Do you always get this excited?" Shafiq asked.

Annja thought about it. "You mean, before I get a chance to look at something that's probably vanished from the minds of men for hundreds or thousands of years?"

Shafiq grinned ruefully.

"Yes," Annja answered emphatically. "I get this excited every time."

With a brief nod at the sails overhead, Shafiq said, "I feel the same way every time I take this ship out onto the salt." He rolled his broad shoulders. "If I ever lose that feeling, I'll have to find something else to do."

Annja leaned on the railing. Out in the distance she spotted luminescent pools of jellyfish colonies. Moonlight kissed the occasional flying fish as they glided across the waves and dropped back into the silvery sea.

"I know what you mean," she replied.

"What do you hope to find?"

"The ship," Annja said. "Answers to questions history has asked but never answered." She turned back to Shafiq, leaned her hips against the boat railing and smiled. "But that's not what I *expect* to find. Only what I hope."

"Then what do you expect?"

"A ship that went down several years ago with interesting trinkets aboard."

"Gold trinkets," Shafiq reminded her.

"Gold doesn't impress me much."

Shafiq laughed. "Then you're among the minority, Miss Creed."

"Call me Annja."

The captain hesitated a moment, then nodded. His cigar glowed. "Gold is the god of many men."

"Gold is a metal," Annja said. "Nothing more. It was prized because of its scarcity, its beauty and because it didn't corrode or rust like most other metals did."

A grin split Shafiq's face. "You take the romance out of our little exploration."

"Platinum was prized more," Annja said. "It was even more scarce. Silver was important, too, but the sea could claim it."

"I know. Saltwater dissolves silver. Gold could be recovered from the sea, but silver—if left down long enough—disappeared."

"Yes."

"Let's hope they were carrying more gold," Shafiq suggested. He paused, as if thinking about whether he should continue the discussion. Then he made up his mind. "You've heard of the continent that so many people believed was lost under the sea out here, haven't you?"

"Lemuria," Annja answered immediately and grinned. "The land of the lemurs. That was a theory that came out in the nineteenth century. Geologists and biologists at the time believed that Madagascar and India were once part of a larger continent. They were trying to explain the biodiversity of the lemur and rock formations found in both areas. So they came up with Lemuria, named for the animal. They believed it was a continent or at least a land bridge that had at one time connected the land masses in this area. Other scientists thought that Lemuria might have extended across the Pacific Ocean and touched Asia and the Americas."

"You choose not to entertain the possibility?" Shafiq seemed amused.

"No. There are some land bridges that can be proven. The Bering Strait, for instance. The land bridge there was called Beringa. It spanned more than a thousand miles and connected Siberia and Alaska, which explains the migration of the North American people."

"Then why couldn't Lemuria have existed?"

"Some land bridges were real," Annja said, "but not most of them. Those theories were developed before the theory of continental drift. And before we learned much about plate tectonics. Sea-floor spreading accounts for a lot of the plant and animal migrations."

"I find such subjects quite fascinating," the captain said.

"Good. I'm glad it's not just me."

Shafiq's cigar glowed again. "We should reach the

dive site tomorrow morning. But before we do, you really should get some sleep."

Annja nodded, stripped off the woolen jacket and tossed the garment back to him.

Shafiq caught the jacket and folded it neatly. "There is another place out here that you might want to keep your mind open about," he said quietly.

"What place?" Annja asked.

"Maybe scientists have given up on the idea of Lemuria, but those who trade in myths and legends still talk about an island that once existed out here. It was known as Kumari Kandam, the Sunken City."

22

Seated in the small galley with her notebook computer, Annja read through the responses her request on the archaeology boards had prompted. Captain Shafiq's information on the subject of Kumari Kandam was limited to tall tales he'd grown up with. Annja had posted a question to find out how much information was readily available.

Even though she'd researched the area where the Shakti excavation had been, she'd only targeted history and legends surrounding that culture.

Kumari Kandam, as it turned out, was a whole new kettle of fish.

The dhow continued slicing through the waves. Most of the crew was on deck making final preparations to the equipment. Shafiq expected to drop anchor within the next two hours. That would put the time at shortly after 1:00 p.m. There would still be plenty of daylight hours to begin searching for the shipwreck.

There were a lot of postings, but not much offered anything significant until one got her attention.

I'll bet you're going to get a lot of responses to your question.

Kumari Kandam is part of reptilian conspiracy. See, there are a number of people—not me—that believe there's actually a lizard people that lived in that time. Or maybe they're still living among us.

Don't know if you're familiar with the Reptoid Conspiracy.

Annja wasn't, but the name alone conjured interest. She slathered blueberry jelly on a fresh-baked biscuit and kept reading.

The Reptoid Conspiracy is basically a belief that if the meteor that had brought about the Ice Age hadn't hit the Earth, the dominant species would have been reptilian, not mammalian.

Some paleontologists believe the Troodon, a small bipedal dinosaur, was destined to become the dominant species instead of Man. From the bone structure of recovered skeletons, paleontologists also believe that the Troodon's features were gradually changing. Their vision was definitely more binocular-like than any other dinosaur at the time. And they had partially opposable thumbs.

Several of those bones have been found in Montana. Others were found in eastern Europe and western China, which definitely puts them in your neck of the woods.

Some people also believe that the Reptoid Conspiracy can be attributed to alien DNA threaded into the dinosaur DNA.

Terrific, Annja thought. That's all I need to add into this. Hints of an alien conspiracy would turn her search for the sunken ship—if it still existed—into a three-ring circus. The pirates would quickly be joined by UFOlogists. She read the rest of the message.

At the heart of the theory, you're dealing with aliens or forgotten dinosaurs that are more human-like than anything we've ever seen.

Anyway, thought you might like to know that. Post what you learn. Curious minds are waiting.

Annja finished her biscuit and sipped her tea. Her mind worked constantly as she tried to sort through the material she'd been given and had turned up. When the Internet connection was interrupted, she adjusted the microsatellite receiver on the table. The connection came back online.

So, she mused, are we looking for a lost ship? Or a lost world?

"DEPLOY THE FISH," Captain Shafiq ordered.

From the stern, Annja stood in the hot sun and watched as one of the crewmen picked up the "fish." The side-sonar drone looked like an old V-2 rocket, or maybe a spaceship from 1950s science-fiction pulp magazines. It was three feet long and had fins at the bottom.

The crewman tossed the drone overboard. The *Casablanca Moon* crept along before the breeze. A side-sonar scan couldn't be done effectively at speed. The towing cable that attached the drone to the side-sonar terminal paid out as the device settled into the water. Buoyant, the drone floated on the surface and immediately trailed the dhow.

A second hull-mounted side-sonar unit was forward of the ship below the waterline.

Lochata, seated under a canvas tarp Shafiq had ordered hung, stared at the terminal. The unit looked like Annja's notebook computer but was thicker. The viewscreen glowed.

"We have a signal," Lochata announced.

Annja retreated back to the shelter of the tarp and felt the immediate change in temperature from the deck to the shade.

The screen showed the computer-assisted representation of the sea floor. The black-and-white image looked a little grainy, but objects, large objects, could be made out easily.

"Nervous?" Lochata asked as she watched the viewscreen.

"A little," Annja admitted.

"You wouldn't be human if you weren't," Shafiq said.

"Good to know I'm human," Annja replied.

Shafiq laughed. "We're treasure hunting. Nothing makes a man feel more alive than that."

"Not treasure hunting," Annja corrected. "Relic hunting."

"Either way," Shafiq replied, "the process is the

same. It's us against the sea. And the sea doesn't like to give up her secrets."

"I know."

"You're sure about the GPS coordinates?"

Annja nodded. She'd taken them with her GPS locator that night and recorded them in her computer, as well as her journal.

"Then it's just a matter of hurry up and wait." Shafiq sighed. "We'll run the search pattern here, then go out farther to sea as we need to. Or another place closer to shore. Wherever you think we might get most lucky."

The search pattern was hard, tedious work. They had to constantly reverse the boat and bring her around to sail back and forth in parallel lines. When they'd covered a big enough area, all carefully charted using the GPS locator on board, they would sail back over the same area in perpendicular lines.

"It's going to be a long haul," Shafiq said. "You might as well make yourself comfortable."

Annja took one of the folding metal chairs from nearby and sat. She hated that she didn't have anything else to do. But sitting and waiting, gazing over Lochata's shoulder at the black-and-white imagery on the side-sonar screen was part of archaeology, too.

"Wait." Annja pushed up from her chair and looked at the viewscreen. But whatever she'd thought she'd seen was gone.

The crew leaned in. Only a few of them were required on deck while the *Casablanca Moon* ran the mile-long search pattern.

All they needed was a trail.

"Did you see something?" asked Paresh, one of the young Indian divers.

"I don't know," Annja said honestly. "I thought I saw a shape there on the ocean floor." She sat at the computer beside Lochata. The terminal posted real-time imagery, but it fed into an external drive on the computer that allowed direct image transfer. External drives were swapped out as needed.

Annja brought up the images, then rewound the copied video loop. Less than a minute later, she had the image.

Everyone except Lochata looked at it. The professor concentrated on the newest images.

On the computer screen, Annja froze the image and studied it. It looked bulbous against the light-colored sand of the sea floor.

"I don't know," Paresh said. "It could just be a brain coral. Perhaps even a rock."

Annja knew that was true. But it was also the most promising shape they'd seen since taking up the hunt. More than that, she needed to *do* something.

"It would take some time to turn the boat around," Shafiq said.

"There's no need." Annja stood. "What's the current depth?"

"One hundred feet," Shafiq answered immediately.

"I'll dive."

A FEW MINUTES LATER, Annja wore scuba gear and flippers. After checking to make certain her regulator was working, she sat on the railing.

Paresh joined her. Shafiq had assigned the young man to accompany her. Annja didn't mind. The buddy system was the easiest way to avoid trouble underwater.

She checked the GPS locator around her wrist, made certain her equipment bag containing a flashlight and underwater camera was attached, shot Shafiq a thumbs-up, took the shark stick one of the crew handed her, then toppled backward over the railing and toward the Indian Ocean.

23

Hands extended to break the surface tension, Annja hit the sea cleanly. The *Casablanca Moon*'s engines became muted as the water filled her ears.

Paresh entered the water a few feet away from her. He slid neatly through the water and righted himself with a couple of practiced flipper movements.

After swimming slowly through the area in a pattern that mimicked the one followed by the *Casablanca Moon,* Annja found the bulbous shape she'd seen on the side-sonar screen. She signaled to Paresh, then pointed.

He nodded and signed that he would go down for a look.

Annja tried to wave him off so she could go instead, but he ignored her and finned down. Irritated but knowing that one of them had to stay up to keep lookout, she finned herself to a holding position ten or fifteen feet above the object.

It wasn't brain coral. Annja knew that straight away because the object was encrusted with concretion. Usually, but not always, concretions formed around iron objects.

That's not good, Annja told herself. The ship you're looking for isn't going to have much iron on it.

Paresh knew what he was doing, though. He swam around the object without touching it. Cracking the concretion underwater often resulted in a pall of black "smoke" pouring out of the break. Even though the concretion took place, so did the iron oxidation.

A quick wave of his hands and movements of his flippers turned Paresh back to look up Annja. He shrugged his shoulders and showed her his empty hands. *What do you want to do?*

Annja held up a finger. *Wait one.*

She finned to the surface, took a quick GPS reading and wrote it down on a waterproof pad that she carried in a watertight pouch wrapped around her left wrist. Then she took a visual sighting on the coastline about three hundred yards away and sank into the water again.

Paresh hadn't waited. He'd evidently decided to follow up on the same idea Annja had. He swam with loose, natural ease farther out to sea. They were beyond the edge of the pattern the *Casablanca Moon* was following.

A quick glance to Annja's right confirmed that the boat was turning around to take up a fresh tack. She kicked her fins and followed Paresh.

GORAKSH GOT HIS FOOD from the window when his number was called and carried it to one of the mismatched tables along the street. The small eatery didn't do much tourist business because it was obviously run-down, but Goraksh liked to eat there.

As he'd grown up, his father had provided him with money to eat out so there was no need to hire someone. For a long time Goraksh had simply thought his father too cheap to hire a woman to cook and clean. Then Goraksh had realized the illegal nature of his father's business and that not hiring a woman to care for him also meant not having someone snooping around his house.

He ate quickly, almost inhaling the food because he was so hungry. The cell phone at his waist hung on him like an anchor. He knew his father was watching the woman archaeologist and that he might be called upon at any moment to carry out a task.

The man who approached Goraksh stood out at once. He was so obviously British, or at least European, in his khakis and light jacket despite the heat. Wraparound sunglasses hid his eyes.

"Hello, Goraksh," the man said. "Mind a little company?" The man sat across the small table from Goraksh without waiting for an answer.

Goraksh blotted his lips on a napkin. "I don't know you."

"But I know you. More than that, I know who your father is."

"Then you should also know how dangerous it is to trouble me." Goraksh grabbed his paper plate and started to get up.

The man reached out and caught Goraksh's forearm. "No trouble," he said. A cold smile curved his lips. "Unless you decide not to listen to me."

Pain pulsed in Goraksh's arm. The man's grip was surprisingly strong. Panic welled in Goraksh. He was certain he could outrun the man, and was even more certain that he could lose him in the twists and turns of the alleys. The man wasn't a local. The slight sunburn showing on his face offered mute testimony to that.

All Goraksh had to do was convince the man to release him, then he could be gone. "All right," he said. "I'll listen." He returned to a seated position.

"Good," the man said, but he didn't release his grip. With his other hand, he placed a file on the table. He left the file closed and reached inside his jacket pocket for his wallet. "I'm James Fleet. A special operative for the International Maritime Bureau."

Goraksh looked at the ID.

"Do you know what that agency is?" the man asked politely.

"Yes." Goraksh was vaguely aware of it and knew it had something to do with law enforcement.

Fleet put the ID away. "A few days ago, an attack took place on a yacht. Maybe you've heard about it." The suggestion was polite and quiet.

"No," Goraksh replied.

A grin split Fleet's face. "Don't ever play poker, Goraksh. You can't lie well enough to bluff."

Goraksh's cheeks flamed in embarrassment, but cold fear drizzled through him. "I could call the police." He was pretty sure the Englishman didn't have any true authority inside the city.

"Go ahead. Ask for Inspector Ranga." Fleet shrugged

good-naturedly. "Perhaps I should call him myself. He might like to be part of this."

Trying to hold on to the panic filling Goraksh was like trying to hold on to sand. Every time he shifted, some of it spilled loose. "What do you want?" he asked in a voice that cracked.

Fleet opened the folder and showed Goraksh the faces of the dead people in the glossy pictures. Goraksh knew them at once. The woman's face looked worse than he remembered. For a moment he thought he was going to be sick.

"Have you seen these people before?" Fleet asked.

"No." Goraksh knew the man could tell he was lying. Embarrassment and fear burned the tips of Goraksh's ears.

"You're sure about that?"

Panic tightened Goraksh's chest and he didn't know if he could breathe. "Yes. I'm sure."

"I don't believe you."

Goraksh didn't know how to respond.

"I know your father had something to do with what happened to the people aboard that yacht," Fleet said. "Inspector Ranga's crime-scene investigators found where the blisters had been affixed to the yacht. That vessel was smuggling, and I'm betting your father knew that."

Barely drawing a breath, Goraksh sat in his chair and waited. Either the man would let him go or he would arrest him. Whichever it was, it would be a relief.

Quietly, Fleet released Goraksh's arm. For a moment Goraksh continued to sit still, certain that this was a trick. He glanced over his shoulder to see if there were any other agents or policemen waiting to take him into custody.

"I know you're going to tell your father about this,"
Fleet said. He smiled. "Feel free to. Let him know I'm
here, and that he's already made one mistake too many.
Now that I'm on to him, he'll make others."

A mistake? Goraksh didn't know what the man was
talking about. But he had tied the people on the yacht
to his father. There was no mistaking that. Someone had
made an error or left a trail.

"You can run along now, Goraksh," Fleet said. "I
just wanted to introduce myself. We'll get to know each
other better over the next few days."

Slowly, Goraksh got up from the table. Only a few
men were at the other tables. All of them were working
men—from the docks, the warehouses, or the stores
and shops in the area—and some of them knew who he
was. Or, at least, they knew who his father was.

Without a word, his heart pounding, Goraksh walked
down the street. He glanced back once and saw Fleet still
sitting at the table as if he didn't have a care in the world.

Goraksh lengthened his stride and wondered what it
all meant. Then he wondered if telling his father about
the encounter would be worse than the encounter itself.
Which would lead to more misery for him? That was
something he didn't know.

AN HOUR INTO THE DIVE, still loosely following Paresh,
Annja was almost lost in the surreal world that existed
under the waves. When she saw the wavering shape off
to the left, barely visible due to the depth and the
distance, she thought at first it was an illusion, something dreamed up by her anxious mind.

Curiosity got the better of her. She finned forward and caught up to Paresh. He didn't know she was there until she reached out and put a hand on his leg.

Paresh pulled his legs in close to his body instantly and flipped over onto his back with his left arm across his body. The reaction might have been humorous if his right hand hadn't been filled with a vicious saw-toothed diver's knife.

Annja fanned her hands and swam back out of his reach until he realized it was her. His cheeks flexed on either side of the regulator and merriment shone in his dark eyes.

He turned his open hands palm up toward her.

Annja pointed two fingers at her own eyes, then back in the direction where she'd seen the motion.

Without waiting to see if he was going to follow, Annja flexed her legs and swam. She slid through the water and checked her compass. The compass still worked underwater. Less than thirty yards in that direction, she detected the movement again. It was along the sea floor. This time she saw what it was—a length of sailcloth trapped amid coral reefs.

Paresh started down immediately.

Annja checked her depth gauge and found that they were already at a depth of 125 feet. Anything deeper for any length of time would require a different mix in their tanks.

She grabbed Paresh's ankle and hampered his movements. His efforts also took her deeper. She could feel the depth now. She knew if they stayed down too long, they'd have to decompress on the way back up. And there was always the possibility of oxygen toxicity affecting the central nervous system.

Annja had a healthy respect for the sea.

Paresh turned around in the water. His irritation showed plainly in his short, jerky movements. He held his palms up.

Annja tapped her depth gauge.

Paresh blinked at her through his face mask, then rolled his arm over and checked his own depth gauge. He held a finger up signaling her to wait. He mimed going down and coming back up, then pointed at his eyes. *Quick look.*

The murk and the diffraction of the water made it hard to judge the depth. Annja knew it could be anywhere from ten to fifty more feet, possibly deeper, because they had no real frame of reference.

Evidently unconcerned, Paresh swam down.

Annja hung back, but her own curiosity pushed at her. She twisted her body and followed.

24

The increased pressure of the ocean closed in around Annja. She focused on the sailcloth and thought about what might be hidden in the folds. Still, she made herself check her dive watch.

No more than two minutes, she told herself.

Paresh reached the folds first. Fish that had taken up temporary residence in the loose canvas fled. A hermit crab scuttled sideways across the ocean floor and stirred up sand.

Annja reached into the equipment bag tied at her waist and took out the underwater camera. As Paresh swam around the perimeter of the canvas, head down to peer under the folds, Annja took several pictures.

A skeleton lay half-submerged in the loose sand. This one was nearly intact, although the left arm was missing. A gold chain encircled the neck, and a ruby pendant lay inside the chest cavity.

Paresh pointed at the necklace but didn't try to take it.

Annja signaled that she could see it, too.

With a quick flick of his fins, Paresh swam along the length of the sailcloth. He used one hand to gently lift the edge from the sand.

After taking a few pictures of the skeleton, the necklace and the sailcloth, Annja swam along the other side. Small fish darted in her vision. Her excitement grew and she knew she was going to have trouble leaving.

Ignoring her, Paresh swam farther out into the open ocean. An eel slid away from him in a susurration of coiled muscle.

Annja put the camera back in the equipment bag and swam after Paresh. When she caught up to him, she pointed toward the surface.

Paresh shook his head, tapped his dive watch and held up three fingers.

Adamantly, Annja shook her head and pointed toward the surface more emphatically.

Looking upset, Paresh nodded. Then he caught himself and pointed farther out.

Carefully, Annja rotated in the sea and stared in the direction Paresh had indicated. There, mired in the sand, lay a ship's timber. It was long and ancient-looking.

Annja's heart raced. A ship's timber meant a ship was nearby.

Unless someone lost one during a storm, Annja told herself. Calm down. Everything is one step at a time.

She tapped Paresh's shoulder. Together, they swam up.

Once she surfaced, Annja took the GPS coordinates and marked them in her book.

Paresh came up only a few feet away. He pushed his face mask up and spit out his regulator. He stared at her and grinned.

"You saw it?" he asked.

"I did," she said, grinning back.

"The ship's timber."

"*A* ship's timber. Keep this in perspective. That could have fallen off a ship just a few years ago," Annja said.

"No." Paresh wiped his face. "That's part of the ship we're looking for. The canvas and the skeleton are there, too."

"We do this one thing at a time," she said.

"But the shipwreck is here," Paresh insisted.

"A piece of it *may* be here. We need to recover that first and see what we have," Annja said, forcing herself to be clinical.

Paresh slapped the water in frustration. "We're wasting time."

"No," Annja said sternly as she faced the man. "We're doing a site recovery. This is how it's done."

"Someone could come by and try to take everything we've found."

"We haven't found everything yet. That's the point." Annja squinted on the bright sunlight beating against the waves. "And if someone tries to dive this spot, we call the coast guard."

Clearly not happy, Paresh looked away and swam on his back.

Annja got her bearings and saw the *Casablanca Moon* coming back toward them for another pass. She waved both hands above her head, then swam for the dhow. She knew how Paresh felt. Impatience chafed at her, too.

AT THE BOAT, Annja caught hold of the ladder a crewman put over the side. The diesel engines throbbed loudly. She hauled herself out of the water.

"You found something," Shafiq said as he offered his hand.

"We did," Annja said. She accepted the towel the captain handed to her and began drying herself. Out of the water now, a momentary chill flashed over her, but the sun quickly started warming her.

Lochata sat at the terminal and stared at the sonar image on the screen.

"We found sailcloth and a body. Close to 150 feet down. We're going to need the Trimix tanks." Annja shrugged out of her scuba gear. One of the crewmen took her tank from her.

"You've worked with Trimix before?" Shafiq asked.

Annja nodded. Trimix was a blended gas made up of oxygen, nitrogen and helium that was intended for deeper dives. The helium supplemented the nitrogen-and-oxygen mix to make it easier to breathe at that depth.

"You know about the problems with using it during shallow dives?" Shafiq asked.

"It's not safe. I know. That's why we'll be using re-breathers for that dive. They adjust the air mix automatically."

"We could work the dive with you."

"No. We'd only lose time." Annja glanced at Paresh. The young diver stood among his peers. From the way he was gesturing at his neck and pointing out to sea, Annja knew he was talking about the necklace. It was the wrong thing to talk about. They weren't there about the gold or the possibility of treasure.

But now it wouldn't leave the minds of the other divers or the crew.

"We'll take one of the lifeboats," Annja said. "We can work the recovery area from it. Let's stick with the search pattern. The tsunami could have scattered that shipwreck all over the ocean floor."

ALL THE WAY BACK to his father's warehouse, Goraksh felt unfriendly eyes on him. He'd looked around as he made his way back, but he hadn't seen any sign of Fleet.

There were many other Europeans and Americans in Kanyakumari these days, though. Then he remembered the IMB agent's comment about Inspector Ranga. Goraksh realized if spies were lurking about, there was no guarantee they were white.

Inside the warehouse, Goraksh felt a little more secure. Then he recalled that the IMB man was actually after his father. Whatever protection remained here was fading quickly.

At the top of the stairs, he knocked on his father's office door.

"Enter."

Goraksh went in and closed the door behind him.

Rajiv Shivaji squinted at his son through a haze of cigarette smoke. His right hand had dropped to the desk drawer and pulled it open. He made no move to reach inside where he kept the big revolver he'd used to kill the woman aboard the yacht.

The image of the woman's slack face in the autopsy photo ran through Goraksh's mind, but it only triggered the memory of the woman's head snapping to the side as his father shot her to save him.

"Something is wrong," Rajiv stated.

Goraksh tried to speak, but his throat was too tight. Instead, he nodded.

"What?" Rajiv pressed.

"There is a man," Goraksh said. "An International Maritime Bureau agent named Fleet. He talked to me at lunch."

Wary interest flickered in Rajiv's dark eyes. "Where is this man?"

"I don't know. I left him there. He had pictures of those people on the yacht."

Rajiv relaxed a little. "What did this man want?"

"He said you've made one mistake too many."

A sneer tightened Rajiv's features. "If I'd made mistakes, he would be here now. He would be talking to me, not you." He shook his head. "He doesn't have anything on me."

"Then why did he show me the pictures of those people aboard the yacht?"

"To get a reaction from you. You're afraid. He was hoping to instill the same fear in me." Rajiv leaned back in his chair and blew smoke at the stained ceiling.

"How did he know about the yacht?" Goraksh asked. "If we'd have been seen, the police would have arrested us." The mere thought of that turned his spine to ice.

"I don't know. It doesn't matter."

"How can it not matter? They're closing in on us."

Rajiv gestured at the chairs in front of his desk. "Sit. We're not going to stay here," Rajiv said. "We can walk away from this warehouse at any time and start up business somewhere else. As long as we're near the ocean, I can stay in business." He stubbed out his cigarette. "Besides, most of our money isn't made through this warehouse. Like our ancestors before us, we take our bounty from the sea."

Knowing his father referred to piracy didn't make Goraksh feel any better. Pirates, when caught, were often given the death penalty.

"You said this man's name was Fleet?" Rajiv asked.

Goraksh nodded.

"Do you know where he's staying?"

"No. I only encountered him minutes ago."

Rajiv sighed. "Perhaps you could have asked him."

That had to have been a joke. Goraksh was certain of that, but his father never smiled.

"The time may have come to leave this place anyway," Rajiv said.

Goraksh tried not to think of his college or the fact that he was being told he would have to give up his life. He wanted to argue the point, but he knew his father wouldn't let him win.

"It appears the woman archaeologist, Annja Creed, might have truly found a shipwreck out there," Rajiv said.

Mention of the woman's name reminded Goraksh of the men he'd sent after Annja Creed's computer. The failure of the men had surprised him. Seeing the shape they'd been in and hearing about the sword she'd used had shocked Goraksh even further.

"If she's found the ship I think she's found, we're going to have to take it from her," Rajiv said.

Disbelief rattled Goraksh. "You—*we*—are being watched. How can you even think about something like that?"

"Because that ship holds the secret to the remnants of Kumari Kandam."

It took Goraksh a moment to realize what his father had said. Even then he couldn't believe he'd heard him right.

"Kumari Kandam? The Sunken City?" Goraksh shook his head. "That's a myth."

"It has become a myth," Rajiv said. "But once Kumari Kandam was real. And somewhere it still exists. That ship may hold the secret to where it is. If they find what I think they will find, we're going to have to relieve them of that treasure."

"How are you going to know what they find?"

Rajiv smiled. "Because I have a man among their crew. I wasn't counting on those idiots you sent after the woman. If they had succeeded, perhaps we might have learned something, but we wouldn't have learned everything."

25

Annja swam through the deep water.

She took several pictures of the sailcloth, the skeleton and the necklace. Two other divers had joined Paresh in the search. All of them had experience in working archaeological dives.

While swimming close to the sea floor, they waved their hands and stirred up the loose silt. The effort removed the loose debris and revealed what lay beneath.

No one tried to remove the artifacts they'd found. That would be the final step after they were certain they'd learned everything they could from the site.

However, the site integrity had already been fouled by the tsunami. Annja's real concern was that they didn't miss anything.

The divers swam in a pattern much like the *Casablanca Moon* sailed. Two divers swam east-west routes while the other two swam north-south.

By the time Annja brought the search to an end, they'd found two more objects covered in concretion. One of them was probably a cannonball, judging from the shape, but there wasn't any way to tell for certain. Another was an amphora. The container was likely used to transport myrrh, oil, wine, olives or grain.

They attached lifting bags from the equipment chest they'd brought down to the objects they wanted to take up to the surface. A blast of air to displace the water inside the lifting bag insured that it would carry to the top where the men in the support boat would reel them in.

Annja took the necklace, seven gold coins of different sizes and a ring they'd found. She dropped them into another bag and tied it onto her belt.

Paresh and one of the other divers bagged the loose bones of the skeleton in a net. Then the lifting bag drifted up with its ghoulish collection. The skull rolled loosely at the bottom of the bag as the shifting water currents moved it.

"I THINK THE JEWELRY is Roman," Lochata announced.

Annja stood under the awning on the *Casablanca Moon*'s main deck. Lanterns powered by the boat's diesel engines threw light over the skeletal remains she'd assembled on the folding conference table. She took video of the bones from head to toe.

Most of the crew sat around them while the boat rocked at anchor. The dark sea spread out in all directions.

"Why do you think it's Roman?" Annja asked.

"Because it fits with what we know of the area and the techniques in use at the time," Lochata answered. "And due to the style of jewelry that's been found. Although I'm no expert, I've cataloged a lot of the finds." She held up the ring. "I believe this profile is Athena's."

"Athena would make it Greek." Annja turned and smiled at Lochata.

"Minerva, then." Lochata put her hands on her hips and stretched her back. "I think I'm more tired than I'd believed possible."

Annja knew Athena was the Greek goddess of wisdom. The Romans had renamed her Minerva.

"The style fits with what we've seen of ancient Roman jewelry," Lochata continued. "It's formed from a hollow hoop that flares out into an oval bezel containing the intaglio of Minerva."

"I'd noticed that." Annja joined the professor and picked up the ring. She ran her fingertip over the intaglio and felt the depression where it had been carefully excised.

"Forgive me, ladies," Shafiq interrupted, "but I'm new at this. What's an intaglio?"

Annja took the ring over to the captain. "Do you have a coin?"

Shafiq produced one.

"Usually jewelry, and other decorations, has an intaglio or cameo styling," Annja said. "A cameo is left when everything else is cut away or an object is deliberately pressed to leave an area higher than the rest of that object. Like a coin." She pointed to the profile that rose from the coin. "This is a cameo." She handed the coin back to him, then handed him the ring.

"The depression in the center of the gemstone?" Shafiq asked. "That's the intaglio?"

"Yes." Annja saw the interest in the man's dark eyes. "The designer cuts the image out of whatever the material is he—or she—is using."

"What's something like this worth?" Shafiq asked.

Annja looked at the captain and wondered what had prompted the question. The crew was all ears.

"A few hundred dollars at most," Annja answered. "Unless you can find a collector willing to pay more."

"And that's hard to do," Shafiq said.

Annja nodded. "Especially since I've sent images of these pieces to the ASI. If they are stolen, the ASI can make a case to get them back."

"So it's not worth a man's time to steal it and risk serious consequences," the captain said.

Understanding dawned within Annja then. The information wasn't for the captain—it was for the crew. She restrained an appreciative grin.

"Exactly," she replied.

Shafiq dropped the ring back into her waiting palm. "Too bad. You'd think we could all be millionaires from something like this."

"Only in the movies, I'm afraid. This is just a dead man's ring. Worth more for historical value than for intrinsic value," Annja said.

"I'D PUT THE RING at third or fourth century," Lochata said, "but the necklace, the pendant at least, is older."

Annja held up the gold-and-glass pendant. The blue glass bead was teardrop-shaped and held fracture lines that caused it to glint in the light. The double-grooved suspension loop was soldered to the horizontal base plate.

"How much older?" Annja asked.

"Probably first century B.C., maybe a little older. I also think it was a piece of something else. Pendants like this are usually used when terminating necklaces or bracelets. This looks like someone broke up a set and pieced them out to family, friends or business acquaintances. Do you see the intaglio on the side of the bead?"

Annja hadn't until the professor had pointed it out. The carved glass showed a centaur wielding a sword.

Carefully, Annja put the glass bead pendant onto a square of soft blue velvet and wrapped it before giving it to Lochata. She did the same with the ring.

"The skeleton is male." Annja led the way over to the bones lying on the conference table.

A few of the sailors hadn't been happy about having the remains on the boat. They'd felt the dead man would bring bad luck. Sailors and superstitions, Annja had found, generally went together like waffles and syrup.

Annja had been able to determine the skeleton's gender primarily from the hips, but there were other indicators.

"He probably stood only a few inches over five feet," Annja said. "Judging from the fact that all of his teeth are here, he was probably in his early twenties when he died. Maybe only his teens. The xiphoid process has calcified with the rest of the bones."

The xiphoid process, the last few inches of the sternum, was one of the last bones that hardened in the human body. Forensic anthropologists often used any present flexibility as an indicator of age.

Lochata touched the skull. "I'd hate to think some mother waited in vain for her young son to come home." Her voice sounded sad.

Annja felt the same way. It was one thing to study bodies that turned up on a dig as indicators of time or situation, but remembering that they'd once been people could be an emotional experience.

"However old he was, he'd had some experience fighting." Annja took her flashlight from her pocket and played it over the skeleton's left arm.

The strong beam brought three score marks into view along his forearm.

"Those are scars from a knife or a sword," Lochata said.

"That's what I'd guess. When I was working Hadrian's Wall in England, we studied skeletons and bones that had been recovered from the battlegrounds. There were a lot of wounds like this," Annja said.

"Swords and knives often got past the small bucklers they carried."

"Those are hard to use."

"You've fought with a sword and buckler?" Lochata looked up at Annja in surprise.

"I've done all kinds of weapons training. I've fought with blades and even jousted."

"From horseback?" Lochata asked.

"There's no other way to do it." Despite her fatigue, Annja couldn't help grinning at the professor's amazement. "Our unknown sailor suffered other damage, as well." She moved the flashlight down to his left thigh bone.

Metal gleamed under the flashlight's beam.

"What's that?" Lochata moved her eyeglasses forward to use one of the lenses as a magnifying glass.

"Unless I miss my guess, that's an arrowhead," Annja said.

"Iron would have rusted away."

"Iron would have rusted away," Annja agreed. "We'll have to take samples to know for sure, but I think it's flint."

"Flint?" Lochata looked at her. "Judging from this wound and the scarring on the arm, our young sailor was at one time a soldier."

"That would be my guess."

"The Romans used metal arrowheads." Lochata rubbed at the gleaming material in the thighbone. "This came from action outside Rome. Perhaps North Africa." She sighed. "I wish there was some way we could talk to the people we find. The stories they could tell would be amazing."

"I know," Annja agreed.

"Can you imagine living so many years ago?"

"I can imagine it," Annja said.

"But to be able to experience it. Wouldn't that be something?"

"It would," Annja agreed, and she thought about the hundreds of years of history Roux and Garin had locked up in their minds that she couldn't get to. She pushed that thought away. There was another story here. She could learn about it.

"With no other apparent injuries done to him," Lochata said, "I'd say he went down with whatever vessel he was on and drowned."

Annja nodded. A chill ghosted across her shoulder

blades and neck. She'd seen drowning victims. It was a horrible way to die.

SHARP RAPPING on the door woke Annja. She blinked against the harsh sunlight slanting through the window beside her hammock and knew immediately that she'd slept longer than she'd intended.

"Who is it?" Annja asked.

"It's Talat, Miss Creed," the young man's voice said. He was Shafiq's quartermaster. "Miss Rai—"

"Professor Rai," Annja corrected automatically.

"She wants you to come quickly," Talat said. "She says they have found the ship."

26

Excitement filled Annja as she stepped up onto the deck. She ran her fingers through her hair and pulled it back into a ponytail.

It was later in the morning than she'd thought. The sun was well up and heated air whirled in from the sea. She checked the surrounding water out of habit, expecting their good fortune to have drawn predators.

Lochata sat hunkered over the side-scan sonar terminal. The crew stood in a knot around the professor and tried to peer over her shoulder.

There was nothing above water to indicate a ship below. The dhow sat at anchor, turned into the wind so she was streamlined.

Shafiq stood nearby with his arms crossed over his chest. His eyes met Annja's and he nodded. A smile stole onto his face.

The crew separated at Annja's approach. When she stood at Lochata's back, she peered at the terminal.

Revealed in ghostly gray and white, the ship's body lay on the sea floor. It had broken into two pieces, either when it went down or when the tsunami had heaved it back to the surface.

"Can you tell what it is?" Annja asked.

Lochata looked over her shoulder, smiled in triumph and shook her head. "No. Not yet. The image is too ill-defined. But you were right about the ship being there, Annja."

"We were lucky," Annja replied. "It could just as easily have not been there."

"But it is."

"How far down is it?"

"One hundred feet," Lochata answered.

Some of the dread in Annja's stomach unknotted at that. The depth was within reach of regular diving equipment without having to worry about decompression problems.

"I'd thought it would be out of range." Annja peered at the faraway coastline.

"According to the charts I have," Shafiq said, "the sea floor is generally twenty feet farther down in this part of the ocean."

"Twenty more feet would put us in a totally different place to do this," Annja said.

"I know," Shafiq replied. "Then you're getting into the realm of commercial diving, and the men you have here aren't trained for that."

"I am," Paresh replied. His eyes remained fixed on the screen.

"Not to mention the fact that we'd need different equipment," Annja said. "But this is doable."

DESPITE THE IMPATIENCE that twisted and turned inside her, Annja delayed her dive until the water eductor was deployed from the *Casablanca Moon*'s stern. The eductor would help her find more than she could on her own.

Paresh, however, wasn't so easily curbed. He paced the dhow's deck and cursed lividly. Everyone ignored him until Shafiq ordered him to help out or shut up.

Also called a water dredge, the eductor was a long metal tube that was used—in effect—like a vacuum hose. Once the seawater started cycling from one end to the other, the flow would continue.

Once it was in place and functioning satisfactorily, Annja suited up with a rebreather and flipped over the dhow's side. The four other divers joined her. In seconds, they were all swimming along the water eductor tube.

The ship took form as a dark shape in the water. It mimicked the sight that had been revealed on the side-sonar terminal.

The sea's salinity had turned the wooden ship black, but it was better preserved than Annja had thought it would be after all the centuries it had remained trapped on the sea floor.

Annja struggled to keep her breathing normal. If she got too excited, she'd burn through her oxygen even with the reclamation and scrubbing properties of the rebreather.

The ship looked like a Roman trade ship. She had a flat deck with a small stern castle and an artemon, a small raking mast at the ship's prow. It had been designed to sail with a beam wind and increase maneuverability. The mainmast had snapped off and only a broken stump remained.

Besides being broken in two, the front piece of the ship was upside down at a three-quarter turn. The stern, all but shattered, sat scattered in a mostly upright position.

Look at it, Annja thought to herself. I found it. I found a shipwreck.

She'd seen pictures and video of several of them over the years, but she'd only been able to guess at what finding something like that might be like.

Her imagination hadn't even been close.

With the shark stick in hand, Annja swam alongside the shipwreck just to get the feel of it. She guessed that it was something over ninety feet long and had a twenty-five-foot beam. Accurate measurements would be hard given the destruction the vessel had suffered.

What did you carry? Annja wondered. From what she'd studied, she knew that the Romans generally brought precious metals to India, gold and silver, to trade for textiles, gemstones, glassware and herbs. For a time, the Indo-Roman trade had been a flourishing business. Then, around 200 B.C., the Roman side of the trade had lapsed, leaving industrious Indian merchants to carry on.

Only a few Roman cargo ships kept plying the waters after that time. No one was certain why the change had occurred. Some historians conjectured that the cost of sending warships to protect the trade vessels had become prohibitive. But some of Rome's inner power struggles had been going on at the time. And there had been unrest along the Black Sea, where several ports had come under attack.

Annja kicked her legs and pulled out her flashlight. When she turned it on, the beam almost disappeared in the murk, but it worked fine inside the ship.

A diver—Paresh, she realized—tried to swim past her into the ship's hold. Annja dropped the flashlight and the shark stick as she grabbed his right leg in one hand and the edge of the hold in the other. She prevented him from entering the broken ship.

Angrily, Paresh turned on her and brushed at her hand.

Annja caught his hand instead, then grabbed his thumb and the edge of his palm in her hands. She twisted the hand into a come-along grip that caused an explosive bleat of pain that she couldn't hear but could see. Paresh spit out his regulator, and a huge air bubble sped toward the surface.

Giving in to the hold, Paresh rolled backward. He kicked at Annja but she deflected the effort easily because the water slowed him down. She released her hold and let him swim away.

The other divers watched.

As Paresh hastily shoved his regulator back into his mouth, Annja slammed her right fist into her left palm and shook her head.

Paresh touched his chest with his hands and exploded them outward with the palms facing him, signing, Why?

Annja took her marker board from her waist and wrote quickly. "Not safe."

Paresh shook his head.

She flipped the board over and wrote on the other side. "Search perimeter first." One of the other divers swam over to Paresh and pulled at his shoulder. With obvious reluctance, Paresh swam off with the other diver.

Annja swam to the sea floor and reclaimed her flashlight and shark stick. She pointed the beam into the ship's hold but didn't try to enter. Sea silt, sailcloth and other debris surrounded chests, chains covered with concretion and amphorae.

Annja suspected the ship had been filled with cargo, but it had been spilled across the ocean bed. Most of it was probably lost. There was no real reason to expect that it would be in the vicinity of the ship. The preservation of the ship's corpse was a fluke. If it hadn't been buried in the sea floor, not much of it would have been left.

Her curiosity at least somewhat satiated, Annja turned her attention back to the job at hand. She had a site to work.

FATIGUED, Annja knew they would have to give up soon. Diving was hard on the body.

Barely moving, she hung over the sea floor. They'd gridded the area in foot-wide squares and numbered them accordingly. That in itself had taken a lot of time.

She moved the mouth of the water eductor over the sea floor. She didn't get it close enough to suction the sand from the bottom because that would have defeated the purpose. Instead, she dragged her palm over the sand and caused flurries to rise that were suctioned away. The work was slow and arduous, but it was thorough.

If they worked the site properly, it could take weeks.

I hope Garin's got deep pockets, Annja thought, because she didn't intend to quit.

They all took turns with the eductor. A moment later, Annja found another amphora. She quickly took pictures with her camera and made a note of the frame numbers and grid location to reference them later.

Amazingly the amphora was intact. There was no way to know what it contained. Annja swam over to the equipment net they'd brought with them and fished out a lifting bag. She tied the lifting bag around the amphora, then filled the bag with air from a spare tank.

The lifting bag struggled to be free, then finally floated up toward the surface. As Annja watched, the bag got bigger. The air inside the lifting bag expanded, and for a moment Annja wondered if she'd overfilled it and it was going to burst.

But the bag made it all the way to the top.

Tired, Annja turned her attention back to the eductor.

Just one more grid, she told herself.

In the next grid she found an oilskin pouch that felt curiously light. As she calmly dusted the sand away from the pouch, she saw the end had been curled under and the whole package had been tightly tied. She took pictures and made notes. Then she plucked the pouch from the sand, tied a lifting bag to it and sent it on its way.

After she picked up her shark stick, she slapped her flashlight against it to make noise. The heads of the other divers swung toward her.

Annja pointed up, then finned up. The others, including Paresh, followed.

By the time Annja reached the surface, all the lifting bags had been gathered by the men Shafiq had assigned to the task. They paddled around in the lifeboats and used fish gaffs to bring them in.

Seeing the items in the lifeboats made Annja feel good about what she'd been doing. This was honest work, the kind of work that she'd been born to do. Nothing like the pop culture pieces she'd been doing for *Chasing History's Monsters*.

Still, she knew if she wasn't doing the show anymore that she'd miss it.

Admit it, she told herself as she tossed her equipment bag and shark stick into the nearest boat, you're no longer purely an archaeologist. You like the attention generated by the show.

She hauled herself out of the water and felt immediately chilled. It was dark enough that the boat crews had to use flashlights to find the lifting bags bobbing on the surface. The ones that were too heavy to take out of the water by manpower, like some of the amphorae, were floated into a net waiting at the *Casablanca Moon*'s port side to be winched up later.

The sailor in her lifeboat pulled up the lifting bag containing the oilskin pouch. Once the bag cleared the water, it deflated immediately.

"Can I have that?" Annja asked as she flicked her flashlight onto the pouch.

The sailor passed it over.

Paresh clung to the other side of the lifeboat and gazed at Annja with hot, dark eyes. His anger stained his features.

Annja chose to ignore him and turned her attention to the bag. She had trouble untying the knots but got them at about the time the last of the lifting bags had been dealt with.

When she opened the pouch, she found a slim book inside. Intrigued, and surprised to find that it was still dry, she pulled out the book.

It was covered in some kind of leather. That marked it instantly as something the Indian cultures would never have made.

When she flipped the book open, she found the script was handwritten, not printed by machine. She couldn't read the language. She wasn't a linguist and she didn't expect to be able to read everything.

What surprised her, though, was the full-color picture on the first page. It was of a beautiful woman. From the top of her head, to her bountiful breasts, to her trim waist, she was human. But from that point on she had the body of a snake covered in iridescent blue, green, burgundy and gray scales.

27

Back aboard the *Casablanca Moon*, Annja worked under the canopy while the cool night air blew around her. She'd changed into jeans, and a long-sleeved blouse.

The book she'd found consumed her interest. She hardly noticed when one of Shafiq's men pressed a bowl of soup into her hands. Hunger prompted her to lift the spoon and eat. But she was careful to keep the food away from the book.

Page after page of the author's brushstrokes met her gaze. The writing was unrelenting in its clarity while still maintaining its secrets.

There were several illustrations, but those tended to be just as mystifying.

Annja finished the soup and pushed the bowl away. She didn't know when it disappeared, but she eventually noticed it was gone.

Finally, her back and shoulders no longer able to take

the continued strain, Annja straightened up. She was surprised to see that most of the crew was belowdecks. A glance at her watch showed her it was after 1:00 a.m. She'd been working for over five hours straight.

"You should get some sleep."

When she turned to track the voice, Annja found Shafiq sitting in a canvas chair with his feet up on the railing.

"It's going to be an early morning tomorrow," Shafiq went on.

"I know. But when I get wired like this, it's hard to sleep. I usually pass out."

Shafiq grinned a little. "You're a beautiful woman, Annja Creed. You should take care not to get old before your time."

Annja didn't know how to take that. Was it merely a compliment, or was something more intended? For the moment she chose to ignore it. At another time, she might have pursued it.

"I thought maybe if you went to sleep, she would, too." Shafiq pointed a forefinger behind Annja.

When she turned, Annja saw that Lochata was still working on the artifacts that had captured her interest. The gold disks spread out on a velvet square in front of the woman told Annja the professor had dedicated her time to the coins that had been recovered.

Lochata had pronounced them Roman based on the numismatic studies she'd done. With all the research done into the Indo-Roman trade history, coins were a good indicator of the time frames.

As if sensing the attention being paid to her, Lochata straightened and turned toward them.

"Were you speaking to me?" she asked.

"Not to you," Shafiq assured her. "About you."

"That can't be good." Lochata picked up an insulated cup of hot tea and walked over to join them. She looked at Annja. "We should get to bed."

"That's what he was suggesting," Annja said.

Shafiq shrugged disarmingly.

"It looks like everyone else has already headed that way." Lochata sipped her tea. "Have you had any luck with that book?"

After a brief, rueful glance at the book, Annja sighed. "I think it's a history, but I can't read it. Apparently the person who authored the book lived on an island," Annja said.

"There are plenty of those in these waters," Shafiq said. "The Andaman Islands in the Bay of Bengal consist of nearly six hundred islands. Not many of them are habitable, of course. Then there are the Lakshadweep and Nicobar Islands in the Arabian Sea and the eastern Indian Ocean respectively. It wouldn't be that unusual."

"Whoever it was, they were obsessed with the *nagas*," Annja said. She shook her empty water bottle, then tossed it into the nearby trash can. "Did you have any luck with the coins?"

"I did," the professor said, nodding. "Based on what I remember, and on the references I had regarding Roman coins found in this area—land and sea—I think the coins were primarily from the fourth and fifth centuries."

"A hundred-year span?" Shafiq asked. "Isn't that unusual?"

"No. Hard currency in those times tended to stay in

circulation. It wasn't like the paper money that gets recycled these days. Money didn't go in and out of political favor, although new rulers were vain enough to want coins struck with their images and symbols of their power." Lochata nodded toward the end of the table where she'd been working. "Those coins are well worn. Traders' coins that had seen a lot of ports."

"Now that you've found the ship, I'm starting to worry. News about this will get back to Kanyakumari and other cities," Shafiq said.

"The coast guard and the Indian navy are within hailing distance," Lochata said.

"If we get the chance to call them," Shafiq said. "I'm just pointing out that it's something to keep in mind. You've found your shipwreck. Maybe now would be the time to sit back and call in reinforcements."

Annja swapped looks with Lochata. Calling in other people was an option. She didn't want to, though. This was her find—*our* find, she corrected herself—and she didn't want to share it with anyone.

"No," Lochata said. "At this point, that's unacceptable. We want to explore more of this site before we turn it over to someone else."

"You don't have to turn it over," Shafiq said.

"Yes," Lochata said, "we would. Once you go to someone bigger, you have no choice but to hand it off."

"All right," Shafiq said. "But we'll keep a weather eye peeled all the same."

"Wake."

Goraksh barely registered the voice before someone

slapped his chest. He blinked his eyes open and glanced at his window. It was still dark outside.

Then the room's bright light flared and hit his eyes with physical force. He groaned, cursed and covered his face with a hand.

"Get up," his father commanded. "We have no time to lose."

Certain his father wouldn't leave him alone until he did as he was ordered, Goraksh sat up on the edge of his bed. He hadn't slept well. Nightmares of the International Maritime Bureau agent pursuing him had all but left him exhausted. The woman's death, her head shattered by his father's pistol, had moved in there, as well.

After a moment, Goraksh peered over his hand. His father was hurriedly packing a suitcase with his clothing. The fear that had lain with Goraksh in his bed returned full measure.

"What's happened?" Goraksh asked.

"A most wondrous thing," his father replied. The first honest smile Goraksh had seen on Rajiv's face in years filled it then. "It appears that Annja Creed's archaeological expedition has borne fruit."

"I don't understand."

Finished packing the suitcase, Rajiv stood and crossed the room. He clapped his son on the shoulders. "She's found a most unusual book."

Goraksh sat quietly and hoped that everything happening was just part of another nightmare.

"Where are we going?" Goraksh asked. "Has the IMB agent—"

"He's done nothing," Rajiv said. He reached into

his pocket and took out a folded printout. "This is about the book."

"What book?" Goraksh felt his father was going too fast.

"The one the woman archaeologist found on the ship. After she had found the *naga* figurines, I'd hoped she would find the ship."

"What ship?" Goraksh felt as if he'd stepped into a play after an intermission.

"The Roman trade ship Sahadeva was sold into captivity aboard."

Like some vague, barely remembered dream, the old tales his father had told him came crashing back into Goraksh's mind. When he'd been a boy, he'd grown up with thoughts of Kumari Kandam dancing in his head. He'd talked about the place where the *nagas* ruled supreme and took care of their people until the jealous water gods pulled the island kingdom below the waves.

Rajiv held forth the printout. It was a bad picture and barely showed the book. But Annja Creed looked stunning as she pored over it.

"I don't understand," Goraksh said hesitantly. He hoped he didn't incur his father's wrath with his ignorance.

Rajiv's face darkened. "Ever since you were a boy, I told you that the blood of kings flows in your veins. I told you that somewhere you had a heritage that you could claim. Now the time has come for you to claim it."

The half-remembered stories bounced around inside Goraksh's skull like popcorn. The wildly improbable stories of a group of the Kumari Kandam people making their way to the coastline of India were just tales his father had told during occasional moments of tenderness.

Those people had tried to find homes among the cities already in existence only to be rejected from place after place. Eventually they found a land of their own somewhere up the Vaigai River and had quietly dropped out of sight of the world.

When Goraksh was a child, the stories had been amazing and wondrous. They'd made him feel special. Even around his father. But in the end, he'd discovered nothing about him was truly special—except for his ability in college. He'd felt certain that would take him anywhere he wanted to go.

But that's over now, he reminded himself.

"Get up," his father commanded. "Get dressed. We have no time to waste. We have a future to secure."

"All right," Goraksh said, but he felt foolish all the same.

Still, he'd never seen his father so animated. That was even more unreal than the stories of Kumari Kandam.

28

With his duffel bag over his shoulder, Fleet paused at the slip beside the coast guard cutter. An armed guard in uniform pinned him with a deck-mounted spotlight.

The man addressed him in Hindi.

With a hand held up to block the intense barrage of light, Fleet said, "I'm James Fleet. Captain Mahendra is expecting me."

"Aye, sir," the guard responded. "Let me get the officer of the watch."

A few minutes later, the officer of the watch appeared at the railing. He was young and rangy, with dark eyes and an easy smile.

"Special Agent Fleet," the officer called down.

"Aye," Fleet replied.

"I'm Lieutenant Rohan."

"Permission to come aboard, sir."

"Permission granted." Rohan gestured at his men. A gangplank was rolled down from the ship.

Fleet shifted his burden across his shoulders. Although the prosthetic foot he wore had stood up under every physical challenge he'd put it through, he still had a tendency to favor it from time to time. It was a challenge climbing the incline. Across level ground it was amazing.

"Special Agent Fleet," Rohan greeted with a crisp salute. "It's a pleasure to have you aboard."

Automatically, Fleet responded with his own salute. "Thank you, Lieutenant. It's a pleasure to be aboard."

"Captain Mahendra expected you earlier."

"I was watching our subject," Fleet replied. He'd been with Ranga observing Rajiv's final preparations to set sail on his own ship. Fleet had remained there until he'd been sure it wasn't a feint.

"Aye. I'd been told he was setting sail," Rohan said.

"You've got men watching him, too?"

Rohan grinned. "Captain Mahendra likes to keep apprised of situations he's involved with."

Fleet started to object. He wanted to point out that he was heading up the stakeout team and that unnecessary duplication of effort resulted in increased probability of being seen.

But he hadn't seen anyone else watching Rajiv. At this point there was no foul. Besides that, he understood the captain's desire to keep abreast of things. Fleet had handled things the same way when he could.

"We've got a sat link on Rajiv's ship?" Fleet asked.

"Aye. Our comm is locked in." Rohan gestured toward the forward hatch. "Captain Mahendra retired early so he could be rested for the voyage."

"Smart man, your captain."

"Aye, sir. He's also assigned you a berth in officers' quarters."

"I'll have to thank him. A ship this size, space is at a premium."

"Aye. It is. But it won't be any hardship. When we put out to sea, we'll be busy."

Fleet felt the ship sliding sideways through the water and gently bumping up against the pier. He'd missed that feeling. For so many years that sensation of being suspended had been a part of his everyday life. His sea legs had come naturally. They still did. Once he stepped onto a boat or a ship, it always felt as if he'd been freed from shackles.

"I'll show you the way to your berth." Rohan started off.

Fleet followed.

"Do you know why Rajiv Shivaji is setting sail? Has the archaeology crew found something?" Rohan asked.

"I've received word the archaeology team made visual confirmation on a Roman galley." Fleet had been in touch with one of the men in Captain Shafiq's crew. That had been a bonus, but it also implied that Rajiv could just as easily have managed the same arrangement.

Rohan nodded as he walked down the narrow, lighted hallway belowdecks. "I admit I have some reservations about leaving those people out there when we know Shivaji is so interested in them."

"What do you think we should do?" Fleet asked.

"Protect them."

"India's coast guard and navy are shorthanded as it is," Fleet pointed out. "Putting a guard over a civilian ship is a waste of manpower."

"Isn't that what we're doing anyway?"

"We're setting a trap," Fleet said. "We don't know for certain that Rajiv will even be interested in those people."

"But the odds are that he will."

"Aye."

"Those people could still get hurt." Rohan pulled open the door to the berth.

"Making sure they don't will make us heroes." Fleet grinned a little, not really feeling the bravado he was talking up. He still remembered the faces of the victims Rajiv and his pirates had left behind.

"And if we can't prevent that?"

"We'll be late. And properly regretful."

"Then let us hope we'll be heroes," Rohan said.

"I will."

The young lieutenant paused for a moment. "Captain Mahendra wanted to make sure you were armed."

"I am." Fleet's pistol and assault rifle were in the duffel bag.

"Do you want to be topside when we set sail?"

Fleet tossed his duffel onto the narrow bed. "I would."

"Then I'll send for you. We'll leave within the hour." Rohan excused himself and departed.

Fleet emptied his bag automatically and stored his clothing in the cubbies. He left the HK 53 assault rifle and SIG-Sauer P-226 pistol on the bed.

As he sat on the bed, he took out his gun kit and

began cleaning both weapons. Like his sea legs, taking care of his tools was second nature.

But he couldn't help thinking how helpless those people on the archaeology ship would be when Rajiv made his move.

He hardened himself to the fear that vibrated inside him. Being late to rescue them wasn't an option.

THE SECOND DAY of the site recovery went much more smoothly. Annja was glad Paresh and the other divers had fallen into the rhythm once they'd realized they were going to find *something*. But the real trick was to find it *all*.

Annja understood their anxiety. She still struggled with it herself. The hardest part of any excavation was proceeding at an appropriate pace. So many things had to be done first. Success wasn't measured just by how many artifacts were uncovered. Equally important were how they'd been found and where they'd been found.

Grids had to be laid to mark the area into searchable, organized quadrants. Videotape had to be shot. Pictures had to be taken. Drawings still had to be made. And all of it had to be documented with written reports.

Normally when a ship went down, it scattered and broke apart. If it hit a reef and tore the bottom out, it sometimes took quite a while before the ship took on enough water to go completely under. During that time the frantic crew could take steps that would allow them to sail on for quite some distance. Cargo, all of it important from a historical sense, could end up scattered for miles.

Annja knew those were challenging maritime archaeology excavations. Sea currents became a major factor in where things eventually ended up.

The Roman trader had gone almost straight to the bottom and been buried until the tsunami brought it forth again. As Annja swam over the top of the ship, she got a chill just thinking about what the last few minutes aboard the ship had been like.

"I THINK I KNOW what took the ship down."

After taking a bite of her cucumber sandwich made Indian-fashion with slices of boiled potato and seasoned with mango chutney, Annja looked up at Lochata. They sat on the *Casablanca Moon* under the tarp.

Annja and the other divers had been underwater for hours. Shafiq had insisted on regulated rest periods. Annja had only put up token resistance. During her rest she busied herself with her journal, notes and inspection of the artifacts.

And she ate.

"According to the history I've looked up regarding this area," Lochata said, "there was a tsunami around 500 B.C. that wiped out much of Kaveripattinam."

"I don't recognize the name," Annja said.

"It's now called Poompuhar," Lochata said.

Annja recalled that name. Poompuhar wasn't far from where they were now. It was almost equidistant from Kanyakumari.

"I called a friend of mine who teaches there," Lochata said. "He researched the tsunami records and found that a tsunami struck the coast in 500 B.C."

"That fits with the time frame we've been able to establish with the coins," Annja said. "But tsunamis don't usually take down ships. The wave displacement out at sea is gentle to ships. Rogue waves are the problems out in deep water."

"I know. But it also depends on the cause of the tsunami. In addition to volcanoes and underwater earthquakes, there are also ruptures in the earth that produce giant gas bubbles. The ship down there could have had the great misfortune to be sitting directly atop such a venting."

Annja thought about that. "The ship is broken in two."

"A violent heave, if strong enough, could have produced that result in an overburdened ship past its prime. You could have gotten such a heave from an erupting gas bubble."

Annja considered that thought while she chewed on her sandwich. It *was* possible, even if highly improbable.

"It's also possible that ship was hit by another vessel— a larger vessel—during the tsunami," Lochata said. "But we have to consider the tsunami was the most likely natural event that occurred that could bring it down."

"You're right. All we have in the end most of the time are theories. We can't know," Annja said.

"Until those times that we can say with certainty what happened," Lochata said.

"I know." Annja sipped some water. "The problem is, we're guessing at answers in history almost as much as we're figuring them out."

"That's part of what makes this job so enjoyable. Remember? I believe you told me that a few days ago when I was complaining about the Shakti dig."

"You're right."

"So we dig," Lochata said, "and we keep digging until we have all the answers or we run out of questions."

ANNJA WAS THUMBING through the pages of the book she'd recovered when the phone rang. She'd just finished shooting images of the pages with her digital camera and transferring them to her notebook computer.

She checked the number on caller ID but didn't recognize it.

"Hello," she answered.

"How is our investment coming?" Garin asked in his deep, rumbling baritone.

"Fine," Annja said as she closed the book.

"You found the ship."

"I did."

"You could have told me. I'm the silent partner in this. Not you."

"We've been busy." Annja reached into the ice chest and pulled out a bottle of flavored water. "Besides, I don't have your number."

"You still have the one that you called me on a few days ago."

"I thought you were getting rid of that one."

"Isn't it amazing how you conveniently phoned that one when you wanted to?"

"Did you call just to harangue me? That's usually Roux's shtick."

"No, I actually called to congratulate you. But I knew you wouldn't believe I'd do something like that without at least delivering a minimum of haranguing."

Weird, Annja thought, but she said, "Thank you."

"I've also been putting some people on this project," Garin went on. "I've got a production studio and a director lined up that will treat this well."

"As long as you're happy," Annja said. She shifted focus. "Have you ever heard of Kumari Kandam?"

"The Isle of Snakes?" Garin asked.

"I haven't heard it called that."

"It was. For a time. And, yes, I have heard of it."

"What have you heard?" Annja asked.

"It sank."

"Terrific. How about *before* it sank?"

Garin laughed at his own wit. "It was a den of iniquity, by all accounts."

"What accounts? Why haven't I seen those accounts?" Annja asked.

"They are stories I was told hundreds of years ago when I was first in India, then again in the nineteenth century when I was amassing a fortune before the opium wars in China. It was a fascinating bit of business. Taking poppy plants from China, raising them in India, processing them aboard ship, then selling the opium back to the Chinese. After the Chinese emperor finally got his troops rallied around him, I plowed most of that fortune into legitimate pharmaceutical companies."

"Less with the I Wanna Be A Millionaire and more with the Isle of Snakes," Annja replied.

"They were cannibals," Garin replied.

29

"Cannibals?" Annja couldn't believe it. If the people of Kumari Kandam had been cannibals, wouldn't that have been mentioned somewhere? she wondered.

"As in, people who ate other people," Garin said. "Yes. That would be them. Luckily they didn't survive in a large number. Just be glad you got there after the island sank. I'd been told they made defensive walls of the bones of their enemies."

"Why were they cannibals?" Annja reached for her journal and opened it to a new page. She took a pen from her pocket and started making notes.

"Because they liked the taste?"

"That doesn't make sense," Annja said. "Nearly all of India is vegetarian. They don't even eat livestock."

"I know. It's sad. They'd rather watch a cow starve to death in the street than feed their own starving children."

Annja frowned. "Okay, Supreme Lord of Sarcasm, maybe we could stick to the cannibalism aspect."

"The stories I remember mentioned that the Kumari Kandam people were xenophobic."

"They didn't care for foreigners. That's natural for island populations. Strangers brought new diseases and new ways of thinking that disrupted the hierarchy in existence there."

"Like Columbus's foray into the New World," Garin said.

"The Vikings were here before him."

"Of course they were. So were the Yoruban people from West Africa. And someone taught the various cultures in South America to build pyramids and do mummification for their dead."

Annja checked the time. She still had a half hour to go before she could dive.

"Columbus brought pestilence to the Native Americans that wiped out millions of them. He also introduced the concept of land ownership. Kings of island empires tend to be territorial. And what better way to deal with strangers than to eat them?" Garin said.

"Not what I would suggest," Annja said.

"Perhaps not, but it was effective. They became known as headhunters, and that's a way of life that's still maintained in various parts of the world. Raiding parties occasionally went to the mainland to bring back trophies."

"I haven't read anything about that."

"Have you heard of the *naga* men?" Garin asked.

Annja thought for a moment. She'd exposed herself

to a lot of information over the past few days. She had good retention. It was in there somewhere.

"Nagaland," she said when she remembered. "In northeastern India. West of Myanmar."

"Exactly. Headhunters still practice there," Garin said.

"Headhunting has been banned."

"Only since the 1990s." Garin chuckled. "Warriors take heads in battle," he said.

"Some warriors," Annja agreed.

"The Celts were headhunters. They nailed skulls to walls and used them to accessorize their horses and chariots. Their culture and bloodline has ties to that part of the world."

Annja knew many historians and anthropologists agreed that the Celts had come out of India at some point.

"Maybe it's something about that part of the world," Garin suggested. "When the United States established economic and military interests in the Philippines in the 1930s, they made an effort to eradicate the practice of headhunting. But a few short years later in World War II, American soldiers took Japanese heads in battle and sent them home to family and friends."

Annja hadn't known that. She made a note to look it up later.

"Of course," Garin said, "the Japanese ate American soldiers. Prison camps during World War II became livestock pens. They cut off limbs of their victims to keep the meat fresh."

A slight sickness twisted Annja's stomach at the thought. It was one thing to view dead bodies and quite another to think about what people might have suffered before their deaths.

"But cannibalism isn't a result of war," Annja said. "It's a way of life."

"Islands have limited resources. Some historians believe the population of Easter Island, after they'd eaten everything else on the island and fished the surrounding waters barren, turned on and ate each other."

"It was three thousand miles to the mainland. And they didn't trust anyone outside the island because slavers had captured over half their people at different times." Annja took a breath. "Aside from the cannibalism, what else do you remember?"

Garin thought for a moment.

Annja thought she heard a young woman's voice in the background. Maybe more than one voice. She felt suddenly annoyed.

"They were snake worshipers," Garin said. "Those figurines you first found could be from the Kumari Kandam people."

"Then why aren't there more traces of them?"

"The island *sank*, Annja. It's hard to recover anything when your world gets swallowed up and dropped to the bottom of a very deep ocean. And there may be more traces of them around than you're aware of. Don't feel badly, because no one else might be aware of them, either. Artifacts often get mislabeled. Or altered. Like the piece of the sword you and the old man found in France."

When Annja had found the last piece of the sword, it had been turned into a necklace.

"Anyway," Garin said, "I have to be going. I've got

empires to pillage. All electronically these days, of course, so the satisfaction isn't the same, but it's something. I just wanted to see what progress you were making."

"How did you find out I'd located the ship?" Annja asked.

"It was on the Internet. *Chasing History's Monsters* is carrying the news, as well."

Terrific, Annja thought. So someone aboard the ship was feeding information to the outside world.

FLEET CAME AWAKE instantly when his name was called. He'd slept in the bed with his back to the wall after the coast guard cutter had left the Kanyakumari harbor. Sleeping like that was more of a doze, and he'd found over the years that he could wake more easily.

Lieutenant Rohan stood in the doorway. The young man's eyes were on Fleet's prosthetic foot standing upright on the floor next to the bed.

"What's going on, Lieutenant?" Fleet asked.

"Captain Mahendra would like you in the control center, sir. It appears Rajiv Shivaji's ship is approaching the archaeological vessel."

"Tell the captain I'll be right there."

"Aye, sir." Rohan left.

Fleet pushed himself to the edge of the bed and began strapping on his foot. When he stood, whole again, he quickly slipped the P-226 into the holster high on his hip and slung the HK 53 over his shoulder. Until he was safe back in port, he wouldn't be without either weapon.

He went up onto the deck and slipped his sunglasses on before he stepped out into the bright, hard sunlight.

The crew noticed him at once. He sensed the edginess about them and realized it was the same feeling he felt inside himself.

An armed guard stood by the door leading to the control room. He acknowledged Fleet with a small nod.

The control room was a few years behind what Fleet had been used to in his special-forces days, but it was jam-packed with electronics. The ship's pilot stood by the rectangular windows and stared out to sea, but he steered as much by the computer systems providing assistance as he did by sight.

Captain Mahendra was in his fifties. A compact man, he nonetheless broadcasted a fierce disposition. He had iron-gray hair, dark skin and a neat mustache.

"Agent Fleet," Mehandra greeted.

"Captain." Fleet came to attention and fired off a quick salute.

Surprised, Mahendra was just a second late returning the gesture. "I was told you had a military career before the IMB," he said.

"Aye, sir. Special boat service.'

"That's a top-notch outfit."

"Thank you, sir."

Mahendra nodded at the navigation computer in front of them. The names of three vessels were clearly written in small letters in English.

Casablanca Moon. Black Swan. Bengal Tiger.

The archaeology vessel. Rajiv's ship. And the coast guard ship.

The *Black Swan* and the *Bengal Tiger* were on convergence courses with the *Casablanca Moon.* The archaeology boat sat idle in the ocean. Fleet couldn't help feeling that it was a sitting duck.

"Well," Mahendra said dryly, "it appears that Shivaji isn't going to muck about with his intentions."

Fleet didn't comment, but he knew that the *Black Swan*'s course wasn't open to conjecture. He knew where she was headed, and she was arriving there with all due speed.

"Shouldn't we stop them?" Rohan asked.

Fleet let the captain answer his young officer's question.

"For what reason, Lieutenant?" Mahendra asked.

"Because Shivaji means to harm that vessel."

"You and I know that," Mahendra said, "but we can't *prove* that." He let out a short breath. "No, I'm afraid that for now we're going to have to be content to be close enough to lend a hand should one be necessary."

Fleet watched and waited as the ship sailed closer to its target.

ANNJA HAD NO WARNING of the danger until it was upon them. She'd allowed the dive team to inspect the interior of the broken pieces of the ship. Paresh had joined her in the prow while the other divers occupied themselves with the stern section.

The water currents rocked the ship every now and again. It reminded Annja that the site was transitory and could change again at any moment.

She helped Paresh roll a heavy chest into a waiting net, then they grabbed the net in one hand each and lifted it from the ship. Annja carried the shark stick in her other hand out of habit.

Outside the prow, they sat the chest down. Covered as it was by concretion, they couldn't open it. The smile on Paresh's face told Annja he was dreaming of gold and gems.

Even if they had been equipped for underwater communication, she wouldn't have bothered to tell him that chests often contained clothing, textiles or trade goods. Pirates and merchants didn't often float around the open sea carrying treasures. Treasure ships, carrying gold and silver from conquered lands, would have been heavily guarded warships, not a merchanter.

As she filled a lifting bag, Annja spotted the hull of another ship closing on the *Casablanca Moon*. She thought it would surely hold up and stay outside the dive buoys, but it didn't.

At that moment, Paresh looked frantic and pointed behind Annja.

When the big, long shadow fell across her, gut-wrenching fear like she'd seldom felt electrified her. She suddenly realized how open the sea was, and how few places there were to hide if someone needed to escape a predator.

Her instincts took over. She pushed herself from the shadow's path just before the rough body of the shark slammed into her.

30

Stunned by the attack, Annja barely stayed conscious. Her breath exploded from her lungs in a spray of bubbles that spun in the ocean and headed toward the surface immediately. The regulator floated in the water before her.

Move! she told herself. If you're going to hang here in the water, you're just a target for that shark or his friends.

She kicked her fins and shot forward. Her left hand captured the regulator and shoved it back into her mouth. She sucked in a deep breath and felt her senses spin as the Trimix cascaded through her system.

The shark stick was falling toward the sea bottom. Annja managed to grab it in her left hand, then angled her body around to track the shark.

The creature was an oceanic whitetip shark. Despite the fear instilled by movies regarding the great white sharks, the oceanic whitetip shark was responsible for more attacks and fatalities against humans than any other shark.

Ten feet long, the shark was a prime example of its species. The pronounced dorsal and pectoral fins—all of them tipped in white, looked oversize on the slim, sleek body.

After missing Annja, the shark became tangled in the loose folds of the net wrapped around the chest she and Paresh had taken from the ship. But that only slowed it for a moment as it fought free and launched itself after Paresh.

Annja swam in pursuit but knew she couldn't hope to equal the shark's speed. The shark's head was small and pointed, but its cruel mouth was open enough to seize Paresh by the calf.

Ink-dark blood spewed into the water. Paresh stopped trying to swim and instead reached for his wounded leg. He tried to hit the shark in the nose, but the water slowed his efforts.

Beyond Paresh and the shark, the other divers swam for the surface.

Annja took a two-handed grip on the shark stick and pressed the end to the predator's head where she judged the brain to be. Then she pressed the trigger and hoped Paresh's leg was clear.

The shark stick jumped in her hands as the heavy .44 Magnum round exploded. With the muzzle pressed against the shark's flesh, there was no room for the expanding gases to vent, which caused even greater recoil. If the shark stick hadn't been so long and as heavy, holding on to it would have been a problem.

The exploding gases vented into the shark's body,

tearing and splitting the flesh even more than the .44 Magnum round did. More blood filled the water.

The whitetip relaxed in death almost instantly.

Annja grabbed hold of the predator's snout and shoved the shark stick into its mouth. She leveraged the mouth open and freed Paresh's leg. There was too much blood, the diver's and the shark's, for Annja to see how bad the wound was.

Movement to her right caught her eye and warned her a split second before another shark shot through the water like a torpedo. She dropped the shark stick and reached for her sword, not even sure if it would appear underwater. Even as she wondered that, the sword filled her hand.

The shark heeled and came back around for her. With the sword gripped in both hands, she slid toward the sea bottom. When the shark missed her by inches, she thrust the sword up and into the predator's midsection. The shark opened up and the water turned dark with blood.

A moment later three other sharks arrived. Paresh was already swimming for the surface. Instead of going after him, the sharks concentrated on consuming their brethren.

While the sharks were occupied with their easy repast, Annja willed the sword away and swam along the sea floor for fifty yards, then angled up. As she swam, she saw that the other ship had come alongside the *Casablanca Moon*.

TENSION FILLED the control room on the *Bengal Tiger*. Fleet stood in the midst of it and watched the satellite feeds showing the activity aboard Rajiv Shivaji's ship and the archaeology vessel.

At the moment there wasn't a problem. The two groups were merely talking. The *Casablanca Moon*'s captain was obviously irate about the casual disregard the new arrival had exhibited for the diving buoy markers floating around the boat.

The big man waved and gestured to the *Black Swan*'s crew. Then two of the *Black Swan*'s crew brandished assault rifles and the captain backed off.

Rajiv approached the confrontation but never stepped into the field of fire. He appeared the paragon of compromise, but Fleet knew the man was anything but.

"Well?" Captain Mahendra asked.

Fleet thought out loud. "If we take them down now, we can get them for weapons violations. Maybe. We can hope drugs are on board. But the only way we're going truly get Shivaji and nail him for the murders I'm investigating is if he has that .357 Magnum he's been so free with."

One of the *Casablanca Moon*'s sailors stepped forward with a pistol. Fleet didn't know why the man had done that. Maybe it was to give a better show of defense, or maybe he recognized Rajiv and guessed at the fate that lay in store for them. Either way, the sailor fired before the captain could reach him.

Rajiv Shivaji staggered back.

Instantly the men with assault rifles opened fire. The bullets chopped into the sailor, drove him back against the railing, then knocked him over the side.

"We're going in now," Captain Mahendra said.

"Of course," Fleet said. As the *Bengal Tiger* got

under way, he hoped that Rajiv Shivaji had the pistol in his possession.

Fleet quit the control room and went out onto the deck. He slipped his sunglasses on to block the bright sun. The sea breeze whipped over him as the cutter surged through the water and the waves began to hammer the hull.

ANNJA HEARD the gunfire while she was still underwater. At first she didn't know what it was and thought maybe something had gone wrong with the eductor. Then, when she surfaced, she saw the men with assault rifles standing on the newly arrived ship and knew that something had gone badly wrong.

She reached for the new ship's stern and pulled herself up the deployed anchor chain. No one noticed her. At the top, she peered over the railing and saw the men standing on the deck. All of them had weapons.

"Where is Annja Creed?" a man bellowed. He wore a Sikh's traditional turban and beard.

Two of Shafiq's sailors pulled Paresh from the water. His leg bled profusely, but it looked as if he would survive his wounds.

The leader of the pirate ship, and Annja knew that's what the ship was, seized a machine pistol from one of his men. He fired an entire clip and the bullets tore at the mast and tarp.

The sailors and Lochata dived for cover.

Angrily, the man thrust the machine pistol back into his underling's hands. "If I do not get an answer," the man threatened, "I'm going to set that ship on fire and kill every person aboard it."

Scared but aware that she had no other choice to make, Annja kicked out of her fins and pulled herself onto the ship.

"I'm here," she said. She loosened the straps on the rebreather and lowered it to the deck as the men turned to face her.

The man drew a pistol from his hip and strode over to her. "Miss Creed," he greeted.

Annja noticed with the slightest bit of satisfaction that she was taller than the man.

He drew back his hand.

Annja reached into the otherwhere for the sword and almost pulled it to her. Even with the sword, though, she knew she'd be hard-pressed to get off the ship without getting killed. She also knew her efforts would probably get everyone else killed, as well. She let it slip from her fingers.

The man took a breath and let it out. He put his hand down without striking her.

"I am Rajiv Shivaji," the man declared proudly. His black eyes gleamed. "The book you found. It belongs to my family."

"That book," Annja said calmly, "has been at the bottom of the Indian Ocean for twenty-five hundred years."

"It has been lost to us for all this time. Now we have it again," Rajiv said.

"What makes it your book?"

"It was stolen by my son's ancestor from those who lived at Kumari Kandam and survived the sinking of the island."

That idea fascinated Annja at once. "There were survivors?"

"Yes. Many of them. They built a hidden city out there in the jungles. That book will tell how we can find it." He turned his head to the side. "Goraksh."

A young man stepped out from the crowd of men. "Yes, Father."

"Go find the book and bring it back to me." Rajiv never took his eyes from Annja's.

The young man hesitated only briefly.

Rajiv Shivaji raised his voice to a bellow. "Who is captain among you?"

"I am," Shafiq called back.

"Let my son pass and give him the book." Rajiv lifted his pistol and pointed it between Annja's eyes. "Or I will kill Miss Creed, then kill you and your crew. Do you understand me?"

"Yes," Shafiq said.

Goraksh crossed a gangplank to the *Casablanca Moon*.

"Do you still have the figurines?" Rajiv asked Annja.

"No," she lied without hesitation.

"I don't believe you." Shivaji's menacing black eyes never blinked. "Do you really want me to kill all those people because you lied to me?"

"No," Annja said.

"Then where are they?"

"In my backpack."

Rajiv grinned and raised his voice. "Bring Miss Creed's backpack, as well."

Annja watched and feared that Shafiq or one of his crew might try to capture the young man.

"They know better," the elder Shivaji said. A slight smile curved his lips. "If they touch him, I will kill them all."

A man burst from the wheelhouse. He pointed north and shouted in Hindi.

Annja turned and looked north. She barely made out the ship just visible on the horizon, but she knew it was coming fast.

"Looks like the cavalry has arrived," Annja commented quietly.

"No," Rajiv Shivaji said. "They only think they have arrived." He spoke in Hindi and the man who'd called out the warning ran back inside the wheelhouse.

Goraksh returned with Annja's backpack and a duffel bag Annja figured contained the book and other things he might have gathered.

"Down on your knees, Miss Creed," Rajiv ordered. He put his hand on her shoulder to force her down.

For a moment Annja thought about how she easily she could have handled him. Putting his hand on her put him well within her grasp. But she knew that his men wouldn't hesitate to kill her, and probably everyone aboard the *Casablanca Moon*. She dropped to her knees.

"This is the coast guard," the bullhorns on the other ship roared as soon as they were close enough. "Put your weapons down and you won't be harmed."

Rajiv gestured to one of his men, who brought him a handheld bullhorn. "Stay back," he said. "If you do not, I will shoot Miss Creed through the head."

Annja had to resist the impulse to strike Rajiv and take the pistol from him.

"Do you understand me?" Rajiv demanded.

"Put your weapons down," the bullhorns repeated.

Rajiv spoke to his men in Hindi. They immediately hauled in the anchor.

"Put your weapons down!" The voice from the coast guard ship sounded harder this time.

From the wheelhouse, the man called Rajiv Shivaji's name. They spoke briefly, then Rajiv glanced to the west where the afternoon sun was headed into evening.

"They've underestimated me, Miss Creed," Rajiv said. "And they're about to pay for that. Stay down or you might get hurt."

Squinting against the harsh light, Annja saw three helicopters fly out of the sun. They vectored in on the ships. In the next instant, rockets jetted from beneath two of them. They screamed through the sky and headed for the coast guard ship.

31

The coast guard team got a split-second's warning before the missiles from the approaching helicopters unleashed fire across the deck. The *Bengal Tiger* shivered beneath Fleet's feet, and he knew she was badly wounded. Flames lapped at the metal decks.

In the same instant, the *Black Swan*'s crew opened fire. Coastguardsmen dropped in their tracks as the bullets took them. A few of them had been blown clear of the ship in the missile blasts.

Fleet's one-eyed vision limited his ability to see his surroundings all at once. He had to keep swiveling his head to take everything in. The first order of business was to save the ship. A naval fighting man had to prevent the sinking of his own ship or take his opponent's.

Captain Mahendra shouted orders to his crew. The *Bengal Tiger* heeled over hard to starboard, and her engines thundered as she sped up.

The two attack helicopters pursued.

Fleet ran to the nearest deck gun as machine-gun bullets strafed the deck around him. A coastguardsman went down just ahead of Fleet.

Without breaking stride, trusting his prosthetic foot, Fleet vaulted the dead man and ran up the steps to the deck gun. It was a 50-caliber weapon and swiveled easily on gimbals.

Operating on instinct and memory, Fleet checked the load, slipped the safety off and took aim at the lead helicopter. Managing all of that was hard as the coast guard ship took evasive maneuvers.

Fleet cursed. Rajiv Shivaji had known the coast guard was on to him. They should have known he'd prepared for them. He'd had the helicopters in the area, either in the water or waiting on a cargo freighter nearby.

Fleet wheeled the gun around and fired by instinct. A line of bullet holes appeared on the side of the helicopter, and he tracked them all the way up to the gas tanks. The helicopter turned into a whirling ball of orange flames and black smoke. A moment later the thunderclap of the detonation reached him. By that time he'd already turned to find the next target.

His attention locked onto the *Black Swan*. Two cargo helicopters hovered over the ship. Rescue baskets were lowered from the bellies of the craft. The crew abandoned the ship and climbed into the baskets.

Part of Fleet admired Rajiv Shivaji's plan. It was bold and daring, and salted with the unexpected. More than that, it was working. All he could hope was that the satellite recon they had on Rajiv would stay with the helicopters. Even so, Rajiv would get away the minute he touched down on land and went into a city.

The second attack helicopter came around suddenly. A missile shot from the pod underneath and came straight at Fleet. He cursed and dived from the deck gun. He was still in midair when the explosion tore the weapon from its moorings.

Heat blistered Fleet's back as he crashed to the deck and struck his head. The air left his lungs. He rolled over and unlimbered the HK 53 assault rifle. He tracked the craft and fired while on his back.

Pieces of the Plexiglas nose fell away. The pilot broke off the attack. In the next instant one of the coast guard cutter's other deck guns opened fire and tore away the tail rotor.

Out of control, the helicopter spun around and around before it fell into the sea. The rotor blades shattered on contact and turned into flying shrapnel.

Fleet stayed down until the carnage was over and the helicopter pulled away and quickly disappeared.

When he glanced back at the *Black Swan,* he saw that the helicopters were just then swinging away with the rescue baskets full.

Fleet pushed himself up and went to see what he could do for the wounded. Captain Mahendra was already there and performing CPR on a man.

"Help me," the captain said when he saw Fleet.

Fleet joined the captain and took over the compression rhythm. "Did you get a chance to call your base and let them know we'd been attacked?"

Mahendra shook his head. "Their first salvo took out all communications."

Fleet glanced over at the comm area. Where the gleaming array of satellite receivers and antennae had been, only tortured metal remained. He kept focused on the compressions and willed the man to live.

But Rajiv Shivaji was getting away.

ANNJA SAT in the bottom of the airlift rescue basket. Constructed with wide spaces between the steel bars that made it up so it could be set down into water to rescue people, the basket looked more like an iron cage with the top missing.

Rajiv kept his pistol pressed against her head. "I apologize, Miss Creed. It seems we're going to need you as a hostage a little while longer."

Annja didn't say anything, but she thought about throwing herself over the side. But now that they were a couple of hundred feet in the air, hitting the ocean surface would be like hitting concrete.

One of the men bound her hands behind her back with disposable cuffs. She settled back against the basket as the helicopter reeled them into its belly.

THE HELICOPTERS MADE one stop for refueling at a merchant ship carrying hidden reservoirs before they reached their final destination.

When they put down again, it was at a private field. Cars waited there for them and within minutes they were whisked away to Thanjavur.

Annja didn't know that until they released her from

the trunk she'd been placed in for the trip. Holes had been drilled into the floor to allow air circulation, which led Annja to determine the car had been used for transporting prisoners before. Before she'd been put in, the young man named Goraksh had gagged her. With her hands tied behind her and limited by the trunk, she had no opportunity to pull her sword.

Her arrival and extraction from the trunk was a welcome blessing. She hadn't recognized the city, but plenty of local businesses had the city's name listed on them. From her studies of Indian maps, Annja knew the city was part of the Tamil Nadu district and was into the interior.

Annja had no clue why they were there. She sat quietly and remained watchful, but her thoughts were on Lochata, Shafiq and the crews of the *Casablanca Moon* and the coast guard cutter. She knew there'd been casualties.

She tried not to think of that. She had to concentrate on her own survival at the moment. She ate what she was given and drank water every time she had an opportunity.

And she waited for a chance to escape.

"YOU'RE A LUCKY MAN, Special Agent Fleet."

While standing outside the emergency room of the Kanyakumari hospital where he'd been treated for superficial scrapes, bruises and burns, Fleet didn't feel lucky. He ached all over.

He looked over at Captain Mahendra, who had also escaped the worst of the attack. "Shivaji got away."

"What is it you Europeans say?" Mahendra asked. "He can run, but he can't hide?"

"That's it." Fleet shook out one of the ibuprofen tablets the doctor had given him in the examination room. He stepped to the watercooler long enough to get a small cup of water and take the tablet. "But this is a big part of the world for him to hide in."

Mahendra sighed. "We're not going to walk away from this."

"I know." Fleet took a breath and let it out.

"Five of my men are dead," the captain said in a quiet voice. "All of them have families that I'm going to have to talk to."

Fleet didn't say anything. He couldn't. He still remembered what it had been like visiting the families of the men he'd lost when he'd been ambushed. That had been the hardest thing he'd ever done. Even harder than the months of painful rehab that had followed as he dealt with the operations and the rehab on his foot.

"Losing men," Fleet said in a quiet voice, "is hard. Trying to make it right to their families is even harder."

"There's no way to make it right," Mahendra said. "Rohan's mother—" His voice broke.

Fleet glanced away and gave the man his space. He hadn't known the young lieutenant had been a casualty until after they'd gotten the fires aboard the *Bengal Tiger* under control.

"His mother asked me to look out for him," Mahendra said.

"You did," Fleet said quietly. "The best that you were able." *You keep telling yourself that every day, too,* he reminded himself. *You don't believe it any more now than you did then.*

"I will talk to her. I will let her know that he was brave until the end."

Fleet leaned against the wall. He was almost too tired to stand. The phantom pains in his missing foot were sharper than they'd ever been.

"What are you going to do?" Mahendra asked.

"Do you mean, am I going to leave?"

Mahendra hesitated, then nodded. "I guess I do."

"I'm not going to leave," Fleet said. "Shivaji broke his cover for a reason. He gave up everything here. According to those people working the shipwreck, he gambled everything on that book that he grabbed from them."

"Do we know what's in that book?"

"No. Professor Lochata believes it's some sort of history," Fleet said.

"Not a journal?"

"No. She said too much care was taken with it. The illustrations were too good."

"What was it a history of?" Mahendra asked.

"The island that sank out in the Indian Ocean." Fleet felt foolish saying that, but he tried to understand what it was all about. Rajiv Shivaji had gambled his life away, and had exposed himself to unrelenting law-enforcement pursuit. He was going to be a wanted man the rest of his days. What could be worth that?

"Kumari Kandam?" Mahendra said.

Fleet thought about the name and thought that it sounded right. He nodded. "You've heard of it?"

"The land of the snake worshipers," Mahendra said.

"Doesn't sound good already," Fleet commented.

"The kingdoms there were supposed to be fabulously wealthy."

"That's not going to do Shivaji much good if it's sitting at the bottom of the ocean. And if it was easy to find, it seems like someone would have found it before now," Fleet said.

"There was also the rumor that not all of those people went down on that island."

"This place is like Atlantis, right?"

"It was also called Mu and Lemuria. It was supposed to be a large continent that bridged India with Africa."

"Nothing like that existed. The continents just broke apart and moved into the positions where they are now." Fleet had caught television shows about that while he was rehabbing and trapped at home. He'd flipped back and forth between the history programs and televised sporting events.

"Perhaps," Mahendra said. "But the Indian Ocean is filled with hundreds of islands. Who is to say that at one time there wasn't one more?"

Fleet had to admit that was true. He scratched his stubbled chin and let out a tense breath. "Do you have a family, Captain?"

"My daughters are grown. Sadly, I lost my wife a few years ago."

"I'm sorry to hear that."

Mahendra smiled thinly. "Oh, it wasn't like that. She decided I was too dedicated to my mistress—the sea. Western ways, including divorce, have been on the rise in this country."

"Sometimes it's mutually satisfying," Fleet said.

"In this case, it was."

"Then let me buy you dinner tonight," Fleet said. "You shouldn't be alone." He paused, then admitted something that he wouldn't have been able to a few years earlier. "Neither of us should be alone."

"I'll take you up on that, Special Agent Fleet. Shall I call you?"

Fleet handed his card over. "My cell is on that."

Mahendra took the card and slipped it into his pocket. "You're not going to be at your hotel?"

Fleet shook his head. "Not when there's work to be done. I'll find Inspector Ranga. We can get a court order to get into Shivaji's warehouse and home. Maybe he left a clue behind."

"You'll let me know?"

"It'll be the first call I make," Fleet assured him.

"Miss Creed hasn't turned up yet," Mahendra said. "By all accounts, she's a very smart woman. Capable, too."

"She's a hostage," Fleet reminded him. "I don't think that's a capable position."

"Still, you have to wonder why Shivaji has continued to hang on to her."

"I've thought about that. Either he's killed her and we just haven't turned up the body, or he's still in-country and in need of a hostage. I'm hoping it's the latter. If she is capable, maybe she can trip Shivaji up somewhere," Fleet said.

32

The basement measured fifteen feet by ten feet and smelled like rotting vegetables. Other than the dirt floor, there was nothing in the room.

Annja has been over every inch of it.

A naked bulb provided illumination behind a security mesh screen mounted in the center of the ceiling. The light barely reached the corners of the room and only weakly dispelled the darkness. The air felt moist. It was easy to imagine that she was breathing in mold with every breath.

Rajiv and his men had taken Annja's backpack and everything from her pockets. They'd even claimed her change. Two of the guards had stayed with her while she'd eaten her meal, then they'd taken her dishes and plastic utensils.

But they didn't know about her sword. It gave her an edge. However, she didn't think she'd be able to escape through the house. Rajiv had too many men there. The only concession was that the room wasn't wired for security monitors.

Stubbornly, Annja pulled the sword into the room with her and rapped the hilt against the walls. The room's dimensions just felt *off*. She worked her way along the stone wall, tapping again and again until she heard a hollow behind one of the walls.

She focused on the sound and allowed herself to think that maybe a tunnel to the surface might exist. The house was old. There was a chance that a tunnel into the basement had originally come from outside. Firewood or coal could have been brought in through a chute.

She used the sword to chip away the mortar between the stones. The length made it unwieldy, but she quickly had a pile of mortar at her feet. The mortar was old and crumbled more easily than she would have thought. Within minutes, she had the first stone free.

She eased the stone out with her fingertips and dropped it to the floor. Even though she peered through the opening, she couldn't see anything. A foul stench came from the void. She continued working.

Only a few minutes later, when she had the opening large enough to stick her head and shoulders inside, she found out why the stench was there. She willed the sword away.

Skeletons, three or possibly four, lay in disarray on the floor inside the void. One of the grinning skulls had a bullet hole through its head. There was no chute or abandoned doorway to the outside. Only a makeshift crypt.

"Ah, I see you found some of my former business associates."

Annja jerked her head from the opening and turned to see Rajiv descending the wooden steps that led up to the main floor. Four armed guards flanked him.

Rajiv looked into the void himself. He pointed to the mortar. "How did you dig that out?"

"It was old. It crumbled easily," Annja said.

Rajiv glared at her. "I come to offer you a proposition, Miss Creed. You're the curious sort. That's what your job is all about. I daresay that's what your life is about."

"You were curious enough to come after that book," Annja said. "You were willing to kill people to get it. And bring the coast guard and law enforcement after you."

"Wouldn't you like to know why?" Rajiv asked.

Annja remained silent, but she was certain that Rajiv knew he had her.

"Come upstairs," Shivaji said, "and I'll tell you how I came to know that book was there and what secrets it contains."

"Anywhere is better than this basement." Annja glanced at the void where the skeletons lay swathed in darkness. "Although I think the company here may be better."

UPSTAIRS, once more in disposable cuffs, Annja sat on an overstuffed couch in a modestly appointed den. Rajiv sat across from her and drank a glass of goat's milk. He'd offered her a choice of drinks and she'd taken water.

"You know about Kumari Kandam?" he asked.

"It was a mythical kingdom out in the Indian Ocean," Annja replied.

"Not mythical," Rajiv said. "Once it was very real. A proud and fierce people lived there for a long time, and they amassed a vast fortune."

"Have you ever noticed how these stories make *vast* get larger with every generation that tells them?" Annja coolly sipped her water.

Rajiv scowled at her.

Since he was still talking to her, Annja guessed there was more at stake than he was willing to admit. He needed something from her.

"As the island began to sink," Rajiv went on, "the people abandoned their city and fled to the mainland." He took the book she'd found in the sunken ship and laid it on the table between them. He opened the book to an illustration of a ship in a harbor repelling boarders.

When she'd first seen the pictures, Annja had thought the ship had been under attack by a hostile force, or that they had been attacking the people on the shore. Looking at the drawing now, she noticed the children. They were on the ship and the shore.

"Most of the people drowned when the sea reclaimed the island," Rajiv said. "But the others—the king and his court—managed to get away."

Of course, Annja thought. The king had all the warriors.

"They took the contents of the treasure room with them," Rajiv continued. He turned more pages.

"Can you read this book?" Annja asked.

"No," he said.

"Then how do you know this story?"

Rajiv hesitated for a moment. "It was given to me by my father-in-law. It's his lineage that carries the blood of the people of Kumari Kandam. He only had a daughter and no son to pass his heritage on to."

"What heritage?"

"Twenty-five hundred years ago, a young man named Sahadeva and his friends found the people of Kumari Kandam, where they built a new city deep within the jungles on the mainland. As you may know, a lot of India's interior is relatively unexplored. There's not enough in the center of the continent to draw settlers there."

The idea of a lost city appealed to Annja. Her interest grew enough that she could almost overlook the fact that she was a prisoner.

"The Kumari Kandam people are fierce warriors," Rajiv said. He flipped to a page showing the picture of a warrior displaying the head of a slain enemy. "They were headhunters. They killed everyone who managed to find their city. Except for Sahadeva."

"What about his friends?" Annja said.

"They killed them, eventually. But not before the king's daughter, Jyotsna, freed him and his friends. She had fallen in love with him."

Annja's immediate reaction was to reject that idea. Then she thought about how attractive an outsider might be to an enclosed society.

"Why were Sahadeva and his friends not killed outright?" she asked.

"The king wanted them to strengthen the bloodline. The community was living too close together."

"They were having genetic defects," Annja said.

Rajiv nodded. "The community was inbred. The children were dying. There was talk of curses and vengeance of the gods."

"They lived on an island community," Annja said. "Capturing or trading for other men and women would have been part of their culture long before the move to the mainland."

"Perhaps. But they were more reclusive. The population along the mainland was strong. And they had much to protect."

"The treasure," Annja said.

"Not just their treasure. Their way of life."

"They believed in human sacrifices." Annja leaned forward and used both hands to turn pages in the book until she reached a graphic depiction of a victim being slain by a man wearing a giant snake's head.

"Yes."

Annja looked at Rajiv. "This man, Sahadeva, escaped."

Shivaji nodded. "He arrived in Kaveripattinam with Jyotsna. There he attempted to sell some of the treasure he'd stolen." He brought out the *naga* figurines. "These were among them. That's how I knew you had found the ship Sahadeva was on when it sank."

Annja looked at the gold figurines as the pieces of Rajiv's story clicked together.

"Sahadeva was betrayed by the merchant he tried to sell the treasure to," Shivaji said. "He was overcome by drugs and sold into slavery aboard a Roman ship. The merchant went aboard the ship, as well, in hopes of finding a better market for the figurines in Rome. While they were at sea, a tsunami struck and took the ship down."

In spite of her situation, Annja couldn't help looking at the book and figurines and wanting to find out the truth of the story.

"How do you know all this?" she asked.

"Because the princess, Jyotsna, lived through the tsunami that wiped out most of the city. She and her children were sold into slavery. She never thought to escape because she was too afraid to brave the jungle by herself and she didn't know the way. But she told the story to her children, and those children told it to their children." Shivaji paused. "Now only Goraksh remains of the bloodline that knows the truth of where they came from."

"What do you want me to do?" Annja asked.

"According to the stories, the treasure is hidden somewhere in the city."

"The book and the story don't mention where it is?" You really shouldn't be entertaining ideas of helping him, Annja told herself. But she knew she was. She couldn't help herself. It was a puzzle and a bit of lost lore and history that she could bring back into the world.

"No," he said.

Annja leaned back. "If I help you, what do I get out of it?"

Rajiv grinned. "Miss Creed, don't misinterpret what's going on here. If you choose not to aid me in this, I'll kill you and wall up your body behind those stones you so conveniently pulled down."

As she looked into the man's eyes, Annja knew Rajiv Shivaji would do exactly as he said he would.

"If you do help me, I'll set you free in the jungle after I've made arrangements to take that treasure out of there. By the time you get back to civilization—if you make it out of the jungle—I'll be long gone."

"How can I believe you?" Annja asked.

"Do you believe that I'll kill you here and now?"

Annja didn't answer.

"If you believe one promise, you can believe both," Rajiv said. "I've nothing left to lose in Kanyakumari. The die is cast. The coast guard and the navy will be looking for me."

Annja stared at the book and the figurines. There's something wrong with you, she thought. You're looking forward to this.

When she glanced back at Rajiv, she said, "All right. Do you know how to get to this city?"

With a smile, Rajiv opened the book to the back end paper. "There is a map, though Sahadeva wasn't skilled in that. But such as it is, with the details of the story and some of the illustrations and information from the book, I think we can find it."

Annja studied the worn piece of parchment. It was a loose map and there was no key. If they found the city, it was going to be a miracle.

"When do we leave?" she asked.

33

They left immediately in the dead of night, which Annja thought was an auspicious time to begin a journey into the unknown. Rajiv had a large group of men and four large trucks that pulled Zodiac boats. The rigid-hulled inflatable boats were capable of deep-water and shallow-water performance. They were perfect for the kind of river terrain Rajiv expected they'd have to traverse.

Annja was kept handcuffed and under watch, so there was no opportunity for escape.

Bide your time, she told herself.

TWO DAYS LATER, after hard traveling through the jungle down earthen roads, trails and finally along the Vaigai River until they couldn't take the trucks any farther, Rajiv gave orders to put the Zodiacs into the water. Within minutes the boats were inflated, the engines mounted, and they were under way.

Annja sat, still cuffed, in the middle of the second boat. The ride was rough because the river flattened out in places and there was a lot of debris to be negotiated.

The branches of teak and mountain ebony trees spanned the river in most places. Below them, even the water looked green. Monkeys and brightly colored birds erupted from the trees at the approach of the Zodiacs.

"Do you want something to eat?" Goraksh sat in front of Annja. During the past few days he hadn't talked to her much, but she'd seen the fearful way he'd acted around his father.

"Please," Annja said. It never hurt to show good manners to the only person she thought might be sympathetic to her situation.

"What would you like?" Goraksh opened one of the food containers.

"Yogurt, an apple and a bottle of water."

Goraksh got the items and passed them over to her.

It was difficult managing the yogurt cup and a spoon, but Annja made it work.

"I'm sorry," Goraksh said.

Annja looked at the young man.

"That you're here. Like this." Goraksh spoke quietly.

"This isn't your fault," Annja said. She spoke quietly, as well, but she knew the men in the boat with them were listening.

"He's my father," Goraksh said. "But I didn't know about all of—" He stopped, unable to go on.

One of the men turned to address Goraksh in Hindi. Scars covered the man's face, mute testimony of his past battles. His voice was hard and flat. He spoke quickly, then turned back to watch the river.

"I'm sorry," Goraksh said. "I can't talk to you. If I do, it'll go harder on you."

"I understand." Annja finished her yogurt in silence.

THEY STOPPED to camp early that evening. When Annja was walked to privacy among the trees for a bathroom break, the scar-faced man and two others accompanied her.

After they were out of sight of the camp, the man grabbed Annja by the hair. He thrust his face into hers and breathed his foul breath over her.

"Don't play games with the boy," the man threatened, "or I'm going to lose you in the river. Do you understand me?"

Annja tried to pull away, but he yanked her back as she'd thought he would. Her body slammed against his and she stood two inches taller than him.

"Don't disrespect me," he said. "Otherwise I could tell Rajiv that you tried to escape."

A tremor of fear passed through Annja. She made herself meet his gaze. "I understand," she said. She tried to look afraid. It was easy to do because she was partially fearful that he would notice she'd slipped the small knife from his hip. She hid the blade in her hands and turned toward the bushes.

"Quickly," the man ordered.

"All right." Annja stepped behind the brush and flipped the knife around. The disposable plastic handcuffs parted easily.

She reached for her sword and found it ready and waiting.

On the other side of the brush, the man slapped at his hip and quickly scanned the ground around him. Then he spoke to the other two men and charged after Annja.

Already in motion, Annja spun behind a tree. Everything had to go down fast because she knew any gunshots or even a cry for help would bring the other men from the boats.

The scar-faced man spotted her. He raised his machine pistol and squeezed the trigger. Bullets raced up the side of the tree and chewed through the bark.

The vibrations echoed at Annja's back. As she whirled around the tree, she dropped into a crouch and gripped the sword in both hands. During the past two days, she'd had a chance to observe the men Rajiv had hired for the expedition. All of them were killers.

Except Goraksh.

She knew none of them would hesitate to track her into the jungle and kill her if they got the chance. You have to make them fear you, she told herself as she came around the other side of the tree.

The scar-faced man saw her at the last moment and tried to bring around the machine pistol. Bullets ripped through the branches and brush above Annja's head as she stepped forward and shoved the sword through the man's heart.

The man froze and looked down at the weapon that pierced him. He let the machine pistol drop from his fingers as he reached for the sword with his other hand.

Shouts echoed from the camp. Monkeys screamed and birds cried out in the trees overhead.

Annja tried to pull the sword free, but the man had a death grip on it. One of the other men fired his weapon. Bullets stitched up the dead man's side and ripped through his scarred face.

Unable to free the sword, Annja grabbed the phone the man carried at his hip. Then she turned and fled into the brush. The men followed her. She held her arms crossed over her face as she ran, but branches still stung her face and neck. A half-buried root tripped her and she sprawled forward. She caught herself on her palms, skidded painfully and pushed herself back to her feet.

Footsteps pounded the earth behind her. Bullets zipped through the growing darkness and dug small craters in the hill ahead of her.

Pausing behind another tree, she reached out for the sword. It filled her hand and she pulled it to her. After checking for her pursuers and discovering they were closing on her position, she willed the sword away again and broke free of cover. She ran away from the Vaigai River. She wasn't worried about getting lost. Finding the river again would be easy enough once she evaded Shivaji's men.

More than that, she knew where they were headed. She intended to follow. But she needed an army at her back.

JAMES FLEET SAT in the office he'd borrowed from Inspector Ranga. He sifted through the digital images Annja Creed had taken. There were a lot of them. He concentrated on the ones of the book that Professor Rai had said was the object of Rajiv Shivaji's assault on the *Casablanca Moon*.

Despite going over the images for two days, he couldn't understand why Rajiv Shivaji would be interested in the book. So far, no one he had found could read it.

His phone rang. His eye flicked to his watch and he logged the time automatically.

"Fleet," he said.

"Where are you?" Captain Mahendra asked.

"At the police station. Where are you?"

"At the base. We just had an interesting phone call from an even more interesting woman."

"Annja Creed." Fleet's pulse sped up. He'd looked through the information Interpol and the New York Police Department had on her. In addition to the work she did in archaeology and on the sensationalist television show, she'd been involved in a number of interesting situations.

If anyone could survive being taken hostage by a pirate and turn the situation around, Fleet had been hoping it would be Annja Creed.

"You've spoiled my surprise," the Captain said.

"She's a pretty amazing woman, all told," Fleet responded. "Where is she?"

"In the interior. Traveling along the Vaigai River."

Fleet got up from the desk and walked over to the wall. Besides a map of the city, Ranga also had one of the country on his wall. Fleet scanned the map and was frustrated when he couldn't immediately find the river.

"Where is that?" Fleet asked.

"As the crow flies, probably a couple of hours."

"What's she doing?"

"She managed to escape Shivaji."

"What's Shivaji doing in the middle of the jungle? He should be running for Europe or Africa at this point. Someplace that won't extradite to India or Britain."

"Believe it or not, he's looking for a lost city," Mahendra said.

"I don't believe it."

"But Annja Creed insists that's what Shivaji is after."

"Is there a lost city out there?" Fleet asked.

"I suggest this to you, my friend. If a lost city was out there and people knew about it, then it wouldn't be so lost, would it?"

"No, it wouldn't. Can you locate her?" Fleet asked.

"The techs were able to lock in on her GPS coordinates through the satellite phone she was using."

"What's going to be done about it?"

"The military is putting together a special-forces team to go in and get her."

Fleet stared hard at the computer. He hated being left out of the action.

"I pointed out that this began as your investigation," Mahendra said. "I told the people who put the insertion together that the International Maritime Bureau needed a representative to look out for their interests. They agreed. If you're willing."

"I'm willing," Fleet said. "Just tell me where I have to be."

ANNJA TRUDGED along the bank of the slow-moving river. Walking in the dark was hard work, and she didn't have a flashlight. Escape wasn't always a convenient thing. It would have been much simpler to escape and be in a city. A cab around the corner, a trip to a bodega, and she would have been set.

Not that she could have used a flashlight anyway. She figured Rajiv and his men were probably still camped somewhere ahead of her. She'd ended up covering more ground than she'd thought.

Now she was tired and hungry. But she couldn't get the idea of the relocated Kumari Kandam city out of her mind. The jungle hid a lot of things. She couldn't help wondering if it still existed. Twenty-five hundred years had passed since anyone had heard of it.

That you know of, she reminded herself. There can be other stories that you haven't come across.

But surely something would be left. Sahadeva had described a large city of carved stone built into the side of a mountain. History had a way of losing things, though, and one simple earthquake could have lost the city again, forever.

She wanted to rest. She wanted a hot meal. And she wanted a bath and a warm bed. Most of all, though, she wanted to see what was at the other end of the map.

The satellite phone vibrated in her hand. She took shelter beside a tree and hunkered down.

"Hello?"

"Miss Creed? I'm Major Anil Patel of the Kumaon Parachute Regiment. I'll be in charge of your rescue this morning."

Despite her fatigue, Annja couldn't help smiling. "Good to hear you, Major. You said parachute regiment?"

"I did. If you'll just hold the line for a moment, my

tech tells me he can locate you." Patel was quiet only for a short time. "Ah. There you are. We're coming in now."

Annja held the line as she looked up at the sky. A plane flew overhead. Tiny black spots dropped from it, then blossomed into parachutes.

"You'll want to be careful here, Major," Annja said. "There's a river close by."

"We see the river. My lads won't be landing in the river, I promise you. We'll be there in a moment."

The phone connection clicked dead.

As she watched the parachutes drop toward the jungle, she wondered if Rajiv or his men could see them, too. If they could, it was going to be a footrace from this point on. She also wondered if they were bringing food. She'd missed dinner by escaping.

34

James Fleet narrowly missed hanging his parachute in a tree. He would have been mortified if that had happened. The paratroopers hit their marks like circus performers even though there was precious little open space in back of the immediate riverbanks. Abundant tree and brush growth covered most of the available terrain.

The inflatable boat cargoes did get caught up in the trees. Without a word, prearranged teams set up a perimeter around the landing area and others went for the equipment.

Out of habit, Fleet pulled the parachute to him and wadded it up. He adjusted his equipment and walked over to Major Patel and the woman.

Annja Creed's pictures hadn't done her justice. Even after days of being a hostage and hours of wandering around in the dark while the rescue effort was arranged, he thought she was strikingly beautiful. More than that, she was composed, as if everything she'd been through were just another day at the office.

Fleet had to admit that after researching her before and after her abduction by Rajiv he had felt a stirring of interest. She was definitely a woman, but one who could obviously take care of herself.

"Do you have any needs, Miss Creed?" the major was asking. He was stocky and powerful-looking.

"Food," Annja replied. "Water. When I escaped, I didn't have time to steal provisions, and foraging in the jungle for edible plants and lizards hasn't become appealing yet."

Patel smiled. "Of course." He reached into his chest pack and took out an energy bar.

"Miss Creed," Fleet said. "I'm James Fleet with the International Maritime Bureau. You might not have heard of the agency, but—"

"I know of the agency," Annja said. She peeled the energy bar open and took a bite. "Who's in charge here? The IMB or the Indian military?"

Fleet smiled in pleasant surprise. "Major Patel is in charge of the rescue operation. I'm here to apprehend Rajiv Shivaji. With the major's assistance, of course."

"Rajiv has a well-armed group of mercenaries with him," Annja said. "You'll need help."

"Right," Fleet said, and decided that she hadn't meant any disrespect. She was just being forthright. "So where is Shivaji?" he asked.

"If I haven't gotten totally lost, he should be just ahead of us."

ANNJA SAT in one of the inflatable boats the Indian paratroopers had brought. They chose not to use the outboard motors and pulled through the river with paddles instead. The men moved in unison, their movements as smooth as clockwork, and they went up the river steadily.

Less than an hour later, with the sun lightening the sky in the east, they came up on Rajiv Shivaji's campsite. The lead boat pulled over to the bank of the river on the side where the campsite was. The other two boats pulled over to the opposite bank while snipers took up positions and other paratroopers got set up to provide covering fire.

After taking cover herself, Annja watched as the shoreline team closed on the campsite. She wore one of the radio earbuds that linked the strike force team. Within minutes she heard the scout leader break radio silence. He spoke in Hindi.

Patel, standing a little ahead of Annja, dropped his rifle from his shoulder and turned to face Annja. "No one is there. Judging from the coals of the fire, they've been gone for a few hours."

"They left before dawn," Fleet said. "They knew Miss Creed got away, and they probably knew she had captured a phone. They chose to put some distance behind them." He sighed. "They might even have seen the plane last night and known we were coming."

"I know where he's going," Annja said. "I saw the map he's following."

Patel looked at the river and the jungle ahead. "Going farther into the interior, against an armed force like that, will be dangerous. It would be better to wait for them to return."

"They might not come this way," Fleet pointed out. "Shivaji may arrange an aerial extraction. Major, I know this is dangerous, but I'd rather not let these people get away if we have a chance to stop them. Shivaji is responsible for several murders."

"My orders were to secure Miss Creed," Patel said. "And to capture Shivaji if I could."

"There's still a chance you could," Fleet said. "And if he gets away, he's just going to kill again."

"Major," Annja said, "Rajiv Shivaji believes he's pursuing a lost treasure out here. If it truly exists, and I don't know if it does, it could be from one of the earliest civilizations ever discovered in this area. It could be a significant find. Not choosing to protect that would be shameful."

Patel grinned beneath his dark-lensed sunglasses. "So by all accounts we get to be heroes here? Either for capturing a wanted killer or for saving a chunk of history? Is that how it is?"

Annja nodded. "I believe so."

"That's how I'll note it in my book," Fleet added.

Patel took a deep breath. "All right. I've always been willing to be a hero. Provided I could survive the experience." He took a map from his Kevlar vest. "Why don't you walk me through our destination, Professor?"

FATIGUE DULLED Goraksh's mind and pulled at his eyelids as he manned a paddle, but fear kept him awake and moving. If his father was right, the police, the coast guard or the military was at their heels. The worst-case scenario was that all of them were in pursuit.

Bright morning sunlight danced on the river water and looked like blazing diamond points. Bird squawks and cries echoed over the area. Monkeys screeched and threw themselves through the trees. Large fish filled the river. Turtles sunned themselves on the banks and occasionally slid into the water.

As he pulled the paddle, Goraksh studied his father's profile. The man looked fierce and proud, like an old-world explorer sailing into unknown lands and ready to lay claim to them for king and country.

The automatic weapons carried by the men in the boats jarred drastically with that image, though.

Minutes later, they rounded a bend of the river that was particularly dense. Short mountains stabbed up from the verdant jungle on the other side. The back of Goraksh's neck prickled hotly. He jerked his head around and surveyed the riverbanks, only then realizing he was searching for whoever he felt was spying on him.

It's nerves, he told himself. That's all. Just nerves. No one is there.

"ARE YOU SURE about this?" Major Patel asked.

As she shaded her eyes against the bright sunlight that penetrated even her sunglasses, Annja studied the wide expanse of the river and the mountains to the right. The river had widened into a large bowl and the water was turgid, almost whirling in one spot. A small tributary fed into the bowl from the northwest.

"Yes." Annja pointed at the tributary. "Look at that

stream. It's not enough to feed this river. That's what tipped off Sahadeva and his friends twenty-five hundred years ago."

Patel stared at the hills in front of them. "Do you really think there's another part of the river that goes underground?"

"To account for this amount of water, yes. That tributary isn't part of the river even though it looks like it on the map," Annja said.

"It's hard to imagine that no one's found this before," Fleet commented.

"There's not much reason to come this far into the interior," Patel said. "Even less to come to this desolate spot. Too much work would have to be done to make the land habitable, and there aren't any natural resources that have been found around here. Transportation costs make getting here too expensive. I'm sure you have areas like that in your country."

"A few," Fleet said.

Patel nodded. "All right, then. Let's find this subterranean leg of the river."

ANNJA EVENTUALLY DISCOVERED the opening. While Patel's men dived in the clear water and tried to examine the riverbanks, she submerged herself, closed her eyes and tried to connect with the current. It took a moment. Her heartbeat rang in her ears. But she felt the whisper pull of the current feeding into the bowl and turned her face toward it.

With the sensation locked in her senses, she followed the current to the southeastern riverbank only a few yards from the way they'd paddled into the bowl. She put her hand into the mass of tree roots and found some of them were broken. A piece of fabric impaled on one of the broken roots fluttered in the current. She snatched it and swam to the surface.

"Over here," she called. She held up the fabric. "Rajiv and his people have already come this way."

The river only went back forty feet without any overhead clearance.

Annja shone the flashlight she carried upward and finally saw the air pocket that formed against the earthen tomb. When she swam up and took a breath, the air was cold and stank of roots and mold.

Fleet surfaced beside her. He spit water and breathed deeply. "I was beginning to think that we weren't going to make it," he admitted. "I was getting ready to turn back, but you just kept going."

"I think the air pocket goes back the rest of the way," Annja said. "We should be okay from here. Probably if we weren't so close to the rainy season the air pocket would be larger."

"Just large enough," Fleet said. "That's all I'm asking."

MANAGING TO GET the boats through the underground passage was harder than swimming through. Thankfully they were designed to run underwater if they had to. Paratroopers piloted them, submerged, through the opening and drove them forward with the engines until the river's roof opened wider and they could once more take their seats in them.

The water and the cave muffled the engine noises,

and Patel felt certain the sound wouldn't carry. Since they couldn't hear anyone else, Patel assumed no one else would hear them.

"How much farther?" Patel asked.

"Are we there yet?" Annja said with a grin.

Patel smiled back. "Ah. I suppose it must sound like that."

"Shouldn't be much farther," Annja said. In truth, though, there was no way to know. Rajiv Shivaji's handed-down story was inconclusive. She based her judgment on the fact that Sahadeva and his young friends wouldn't have gone too far into the subterranean river cave without knowing what lay ahead. And they hadn't been as well prepared as the paratrooper team of Rajiv's mercenaries.

It was farther than Annja had believed. At least half a mile passed before they reached the other underground river entrance. Annja, Fleet and two other paratroopers who were strong swimmers swam through the opening.

They came up in another bowl that was slightly larger than the one they'd entered. The entrance to this bowl was on the bottom, thirty feet from the surface. It was easy to understand how it hadn't been found because the river would have had to have gone dry before anyone could see it.

Once they were certain of the way, Annja and the others returned to the subterranean section of the river. Swimming with the current made the trip easier.

They piled into the submerged boats and held on as they powered through the opening. On the other side, the boat pilots increased the buoyancy of the boats and they surfaced in seconds.

Annja stared around at the jungle and the slow-moving river that wound through it to the east. The vegetation was lush and thick.

"Well," Patel said brightly, "that was a bit of fun. Throw in the *Pirates of the Caribbean* and we might have a tourist attraction."

Annja smiled at him.

"So where's this lost city?" the major asked.

Before Annja could answer, gunshots sounded from upriver. There were only a few at first, then a multitude as assault rifles opened up on full-auto.

35

Cloaked in the overhang of the craggy faces of the mountains on the south side of the river, the city wasn't immediately visible. Trees, brush and vines had overgrown it for years.

Only the remnants of the docks carved out of the stone at the foothills of the mountains remained.

"There," Rajiv said.

Goraksh heard the excitement tightening his father's voice. In contrast, Goraksh's stomach churned violently. The stories he'd been told since childhood came back to him full force. The people of Kumari Kandam had been fierce warriors. They'd claimed the heads of opposing warriors in battles.

Twenty-five hundred years later, Goraksh knew it didn't matter that some lost drop of their blood flowed in his veins. He was an outsider. They would only recognize him as an enemy.

At his father's order, the Zodiacs headed toward shore. They kept the engines off and used the paddles. It wasn't just to prevent anyone who might be following them from tracking them. It was so anyone who might still live in the city wouldn't hear them coming.

The sensation of being watched prickled the back of Goraksh's neck again. He stared at the morass of tumbled rock and overgrown vegetation.

Twenty-five hundred years later, he thought, could *anything* remain of those people?

Goraksh honestly didn't think so. Probably the city had been found shortly after Sahadeva had been betrayed and sold to the Roman trade ship. The treasure— if any had ever really existed—had to have been taken.

The Zodiac bounced when it made contact with the stone dock. The man seated in the prow stepped out and secured the mooring lines to the stone cleats that someone had carved there long in the past.

Goraksh picked up the assault rifle his father had assigned to him and went forward with the group.

In back of the docks, the city proper took shape. Despite the overgrowth and fallen rock, Goraksh could see what remained of the plaza. The ground was covered in flagstones that might have been mortared at one time to keep the grass from growing. Now vegetation grew freely between the flagstones and tilted them at treacherous angles.

The courtyard was almost the size of a soccer field. The length lay along the river, and the width butted into the caves.

Stone statues of *naga*s stood outside the largest and

most ornate door that had been carved in the mountain-
side. The two male *naga*s stood curled on their serpentine lower bodies. They held shields in one hand and
spears in the other. Helmets covered their faces.

"We must find the throne room," Rajiv said. "That's
where the treasure is supposed to be kept." His voice
sounded low and hollow in the quietness that filled the city.

Goraksh followed his father through the arches.
Inside, he found the mountain had been gutted. The area
had once been a marketplace. Shattered carts and stalls
lay in various stages of disrepair. A cistern filled with
vegetation growth that had crept in from outside, either
through roots seeking a water source or seeds riding air
currents through the doorways, occupied the center of
the area. Three twenty-foot teak trees took precedence.

Buildings had been sculpted out of the mountain, too.
Once, Goraksh thought, the city must have been beautiful and elegant. Now it was nothing more than a ruin.

"Search the buildings," his father ordered. "Find the
throne room."

Goraksh followed his father across the plaza and into
a building. The area was so dark inside that they had to
use flashlights to light the way.

The building had been a shop, filled with shelves and
products that had turned to dust over the years. It was
filled with shadows that seemed to twist to avoid the light.

Then one of the shadows stepped forward and attacked.

"Look out!" Goraksh yelled in warning as he tried
to bring up his assault rifle. His flashlight beam, for just
an instant, fell across the face of his attacker.

The man was misshapen, as if all his parts belonged

to different individuals. Shaggy white hair and beard plunged past his shoulders. Skin white as paper glowed under the flashlight's beam. He was naked and bestial. His eyes glowed red in the flashlight beam.

He carried a thick club. Goraksh saw it for just a moment before it crashed into the side of his head. He fell to the floor and tried to hang on to his senses as gunfire broke out around him. Then a black pit seemed to open up in front of him and he fell into it.

AT THE SOUND of gunfire, Patel gave the order to start the outboard motors. In seconds the inflatables shot across the river surface. Birds in the trees crowding the riverbanks took flight in bursts of bright colors.

Annja held on as the boat jarred and jerked. Her sunglasses blunted the wind.

"Miss Creed," Fleet called from beside her.

She looked at him. "Annja," she said.

He held up a gun belt containing a semiautomatic pistol. "Do you know how to use one of these?"

For a moment, she hesitated. She didn't like guns. She knew how to use them, and she'd used them before, but always in the heat of the moment. To deliberately strap one on and accept the knowledge that she was going to have to do harm to someone else with one of them wasn't a comfortable thought.

Still, not helping defend them if it came to that wasn't acceptable.

"Yes," she answered. "I know how to handle small arms."

Fleet grinned and the effort pulled at the tiny, fine

scars around his eye. "You know the proper nomenclature. Who taught you?"

"Ex-SAS soldiers while I was working a dig at Hadrian's Wall," Annja replied.

"Well, I can't fault you for your training, then." Fleet handed her the pistol and passed over extra magazines.

"Thank you." Annja pulled the gun belt around her hips and tested the fit of the pistol in the holster. It pulled free easily.

Gunshots continued to sound ahead of them. As the boats raced forward, the sounds grew louder.

GORAKSH WOKE to someone pulling on his arm. He tasted blood in his mouth. When he blinked his eyes open, he saw that his father had hold of him. His father's flashlight shone on a horrible tableau.

The man who'd struck him sat slumped against the nearby wall. Bullets had wrecked his face and exploded the back of his head.

"You shot him?" Goraksh asked.

"Yes," his father replied. "I thought he had killed you."

Goraksh moved his head gingerly, afraid that it would topple from his shoulders if he wasn't careful.

"Can you walk?"

"Yes. I think so." Goraksh's vision was still slightly double. The lack of light didn't help. He held on to his father's arm as Rajiv pulled him to his feet. His father's strength surprised him.

When he bent down to retrieve his flashlight, Goraksh almost passed out. He curled his fingers around the flashlight and brought it up. He shone it around the room and saw that some of his father's mercenaries had been killed. The survivors looked frightened.

Several albino men and women lay sprawled in the room.

"There were more of them," Goraksh said.

"Yes. Things got complicated, but we appear to have fought our way free of them for the moment. More of them are still out there. They won't remain afraid forever. Maybe not even for long," Rajiv said.

"Who are they?"

"I don't know." Rajiv picked Goraksh's weapon from the stone floor and handed it to him. "Be prepared."

"Perhaps we should return to the boats," Goraksh suggested.

"And go back empty-handed? *Never!*"

The fierceness in his father's voice let Goraksh know his father would never be dissuaded from his goal.

"We have people following us," Rajiv said. "If we don't claim the treasure now, or kill those who trail us, we may never get another chance."

Goraksh shone his light over the dark areas. He wanted to plead with his father, but he knew even his best efforts would fall on deaf ears.

Rajiv rallied his men and told them that the gold was too close to leave behind. There was only a slight hesitation in their ranks, but the lure of gold was too much.

"These are primitives," Rajiv shouted. He kicked one of the prone bodies. "They caught us by surprise, but now we know they're here. Now we stay together."

Outside the sound of racing outboard motors grew louder. Goraksh listened with a sinking heart. Going back or going forward was the same. Either direction promised uncertainty and potential death. His father would never surrender.

The group lurched into motion.

As THEY CAME around the bend, Annja saw the pirates scattered across the stone pier that had been carved out of the earth on the riverbank. The men took positions and opened fire on the paratrooper boats at once. Minigeysers shot up in the water where the bullets slapped the river.

One of the paratroopers fell out of the boat beside Annja. Without a second thought, she dived from the boat and went headfirst into the river. The water was clear enough for her to see easily.

Blood streamed from the paratrooper's head. Annja couldn't tell how badly he'd been hit, but she hoped she wasn't chasing a dead man. Below the man, several boats littered the river bottom. Her heart leaped at the potential for a treasure trove of archaeological finds, but she focused on saving the drowning man.

She darted behind him, caught him up under the arms in a lifeguard carry and swam for the surface. When she was up top again, she made sure his face was in the open air.

He was breathing.

All around her, the battle had taken shape. On his knees, Fleet fired his assault rifle and raked the pier with rounds. Rock fell from the carved arches over the entryways behind the dock area.

Annja hated watching the destruction taking place. The docks were potentially an archaeological prize.

Please don't destroy everything, she thought desperately. With her arm under the unconscious man's chin, she kicked for the dock's edge.

36

The inflatables butted expertly up against the docks and rebounded gently before a man in each boat tied the mooring ropes to cleats. The mercenaries were firmly entrenched at the main archway and were holding their positions with ease.

Annja saw a spherical object sail through the air just as she grabbed hold of the pier. She didn't realize what the object was until it exploded and tossed the bodies of dead men in all directions.

A section of the archway toppled to the flagstones and shattered.

Annja nearly wept at the sight of everything that had been lost in just that split second. Even from her distance she could see that the archway had been carved with figures and symbols. There was no way of knowing what had been destroyed.

"Here. Let me have him."

Dazed, Annja looked up and saw Fleet squatting on the pier in front of her. He reached down and caught hold of the unconscious man's Kevlar vest and pulled him out of the water. Annja helped push him onto the pier behind a jumble of rocks that had probably been used as ballast in years gone by.

Patel's shock troops moved forward in covering positions as Annja pulled herself onto the pier. She knelt beside the fallen man and checked the jagged tear along his temple.

"He's still alive," Fleet growled. "He should stay that way if he doesn't bleed out."

Annja retreated to one of the boats and grabbed a med kit. Bullets chased her, but Patel's men took down at least two of the mercenaries. Annja saw their bodies fall.

Back at the wounded man, Annja knelt and pulled out a pressure bandage. The ballast rocks covered them for the most part.

Fleet put his assault rifle aside and lifted the man's head.

Annja wrapped a pressure bandage around the head to slow and, she hoped, stop the bleeding.

"I've known men like this nearly my whole life," Fleet said. "Even though I've never met this man and he's from a different country, we're brothers."

Annja saw the fierceness in the man's gaze. She knew that he cared enough to save, as well as to kill to protect. Those two things didn't always go together. She finished the bandage and gave the man a shot of morphine. If they didn't hold their own against

Rajiv's mercenaries, he wouldn't stand a chance, but she didn't know what kind of damage the man had sustained.

"Has anyone called in air support?" Annja asked as she washed her hands in the river.

Fleet grinned. "Small arms and air support. Are you sure you're an archaeologist?"

"First and foremost," Annja said. "Last, too. It's the times in between that give me trouble."

"Well, let's see if we can get out of here and I'll stand you to a dinner," Fleet said.

Annja couldn't believe he was asking something so innocuous in the middle of a running gun battle. Even more, she couldn't believe it when she said, "All right."

Fleet nodded forward. "Looks like we're going in."

When Annja looked up, she saw that Patel's men had stepped through the arches. She followed Fleet. Seconds later, they saw the first of the albino men and women mixed in with the dead and wounded mercenaries.

Fleet cursed. "Look at their hands," he said.

Each of the dead men's hands was festooned with fingers. There were as many as eight and nine jutting off in all directions. Most of them would have had problems doing anything that involved fine motor skills. Other men showed burn scarring where they'd probably amputated fingers that had offended them most.

"Postaxial polydactyly," Annja said.

"Greek to me," Fleet said.

She knelt and picked up one of the dead men's hands. She touched the fingers and felt for the bone within. Most of them were pliable.

"Boneless," she said. She let the hand drop. "Post-axial polydactyly is a genetic disorder caused most prevalently by inbreeding. Several members of the royal families in Europe exhibited similar problems."

"None of the women have the extra fingers," one of the paratroopers said.

Annja shook her head. "They won't. Polydactyly is tied to the male sex. The albinism can be a trait caused by inbreeding, too, but not necessarily. That genetic defect also occurs all on its own."

"These people look like savages," one of the paratroopers said. "They don't even clothe themselves."

"Diminished mental capacity can be another problem of inbreeding in closed societies," Annja agreed. It made her sad to think the people that had escaped Kumari Kandam had come to this. But they had chosen to exile themselves from the world.

They were cannibals, Annja reminded herself. That'll exclude you from a lot of people.

"They move like animals," another paratrooper said. "They have a *gait,* not a walk."

"They may not be far removed from animals at this point," Annja said. She tried not to think about the possibility of children among these people, but she knew it existed. "There may be children."

Another sporadic burst of gunfire sounded ahead of them. Without a word, they gathered their gear and headed deeper into the lost city.

HEAD STILL POUNDING from the blow he'd received, Goraksh reloaded his machine pistol and breathed in great draughts of air that tasted like blood and cordite. He didn't know how many of the snake people he had shot down. He didn't want to think about it.

"The army is pursuing us," one of his father's mercenaries said.

Rajiv drank from a canteen, then hung it once more at his hip. "Then we should not be here when they arrive." He started forward again. "Hopefully those animals will attack them, as well."

The men fell in line behind him. Goraksh followed, but he doubted their only thoughts were of gold.

Only a short distance on, they came to a large amphitheater carved from solid rock. Flashlight beams bounced around the room in crazy arcs. There was enough seating, carved in elliptical rings around the floor at the bottom, for thousands of people.

The room smelled like an animal's lair. The thick scent coiled in Goraksh's nostrils and mired his lungs. He had to work hard to breathe. He turned as the feeling of being watched plagued him again.

His flashlight beam skidded across the stone walls and revealed glimpses of the beautiful mosaics that had been laid across the surfaces. Chief among the images were the *nagas,* half people and half snakes with savage faces and weapons. Here and there, skulls mounted in the wall glared at Goraksh with hollow eyes.

Something moved beside him.

Panicked, Goraksh swiveled to the right and played his beam in that direction. A section of the wall spun open and two of the beast-men stood there with swords. They attacked without warning.

Goraksh managed to scream—once—then he was drawn into the blackness.

FLEET KEPT abreast of Patel as they jogged through the passage and followed the two scouts. The men kept a running dialogue through the headsets, but it was all in Hindu and he didn't understand a word of it.

Three doorways took shape in the tunnel ahead of them, all of them brought out of the darkness by light coming from inside the room. The light jumped and moved, then gunfire erupted.

Patel barked an order and followed it immediately in English. "Down!"

The paratrooper team went down on their bellies at once and shouldered their rifles. Fleet did the same, but glanced back to check on Annja. He was pleased to see that she had gone to ground, as well. She even had her pistol in hand.

Good girl, Fleet thought automatically. He was just as sure a heartbeat later that she would have taken offense if he'd made the mistake of saying that aloud. She didn't seem the type to need praise—or condescension.

Hoarse screams and frightened yells reverberated in the room ahead of them. Slowly the lights winked out.

Fleet glanced at Patel. He knew from experience that the man was thinking that the mercenaries weren't his men and he wasn't responsible for them, but all the same it was hard to stay there and listen to them dying.

Patel spoke again and waved his fist. Fleet was already on his feet before the English order came to advance. The paratroopers went forward and fell into three groups spread along the doors. Other men watched both ends of the tunnel.

At Patel's order, the men threw flares into the room. Bright ruby light exploded to life with hollow bangs inside the room. They waited only a moment, then moved inside the way they'd been trained. They fanned out and took up covering positions.

Fleet knew automatically that there weren't as many mercenaries in the room as there should have been. Only a small knot of them were trying to fight back from the center of the room. They huddled there among a morass of rocks and skeletons. Blood slicked the floor in several places.

Annja dropped down beside him and scanned the room. Before the first flares had time to die out, Patel ordered his men to throw more. A moment later, they tossed grenades into the pile of men at the bottom of the amphitheater. The explosions blew the remaining fight out of the rest of them.

After the surviving mercenaries—six of them in all—threw down their arms and surrendered, Patel's men went down and secured them. Fleet went down, as well, but the whole time he felt as if he was being watched.

Rajiv Shivaji wasn't among the mercenaries.

Patel questioned the men quickly and roughly. "Shivaji isn't here," he told Fleet. "They say the beast-men got him."

Fleet regarded the several albino bodies lying on the ground. "Then where is he?"

"These men say Shivaji and some of the others went

into the walls after the beast-men. They captured Shivaji's son."

"The walls?" Annja said. She swung her light onto the walls immediately.

Fleet caught sight of the images set in the stone. He played his own light over the pictures of snakes wrapping opponents in battle.

"These…things we've seen couldn't have made these," Fleet said.

"Not them," Annja agreed. "Their ancestors. Before the inbreeding became so severe. Until that time, they were probably pretty normal." She moved the flashlight and hesitated over the places where bullets had scarred the pictures.

"This place is a really big deal, isn't it?" Fleet couldn't help asking.

Annja nodded. Her tone was reverent. "It is." A moment later, she moved toward a statue near the wall.

The statue showed a *naga* warrior curled up on his tail while holding a spear and a shield. His inhuman face looked almost demonic, an eerie mix between snake and human that left the finished product with a forked tongue and long fangs.

"Obviously they had a thing about warriors," Fleet said.

"Most cultures do," Annja said. "It's all about prowess. About being able to take what you want and the ability to protect yourself. Those are the same components behind most sports competitions."

"You're probably never at a loss for conversation, are you, Professor?" Fleet asked.

Annja smiled at Fleet then, and he thought the effort

was dazzling despite the wet hair and dirt stains on her face. "Depends on the listener. Some people think I'm boring and I'm not a professor, I just talk like one."

Fleet laughed. "So what is this place?"

"The amphitheater?"

Fleet nodded.

"Probably where they came to discuss community issues. I'm pretty certain that's a sacrificial altar under the debris. That's why there are so many skeletons here. Even after their minds started clouding, they still returned here to follow the old ways to the best of their ability. If you sort through those bones, you'll find that a lot of them belong to animals."

"No, thanks. I've seen quite enough of that."

Annja stepped forward and examined the floor. Patterns of disturbance showed in the dust. "I think there's a door here." She pressed against the wall.

A click sounded that carried in the hollow silence. Almost immediately a section of the wall spun out at a ninety-degree angle and presented an opening into a well of darkness beyond.

"Charming," Fleet said sarcastically.

"They went this way, though," Annja said.

"How do you know?"

Annja shone her flashlight on the blood trail that led inside. "Someone dragged a body through here." She followed the light through the opening before Fleet knew she was going into it.

The wall section slammed shut behind her.

Fleet stepped forward and tried to operate the wall.

It would not budge. He shouted Annja's name. There was no answer.

Feeling desperate, he turned to Patel. "We need to take this wall down."

37

On the other side of the wall, Annja heard the movement above her too late to do anything more than duck. Something whistled by her head. She turned and dropped into a fighting crouch. After she shifted hands with the flashlight, she reached for the sword and pulled it into the passage with her.

The flashlight beam showed enough of her immediate surroundings to let her know that she was standing in a tunnel. Dust, inches thick, coated the floor and softened her footprints and those of her attackers.

The beast-men advanced on her from another doorway that opened in the wall behind the first one. Annja guessed that the mountain was honeycombed with tunnels, all of them running in different directions and serving different purposes.

Some of the tunnels had doubtlessly allowed leaders of the ceremonial gatherings inside the amphitheater to enter and depart the assembly without walking through the main halls. Others might have been there for escape or to ferry food and drink on festive occasions.

No city, not even Rome in all its antiquity by present standards, ran without its subterranean, other self.

As the beast-men advanced, Annja realized that light came from inside the other room. They probably couldn't see in the dark, either. That inability made her feel better about her situation. Even the narrow confines of the tunnel worked in her favor.

"Stay back," she warned as she gave ground slowly. She kept the sword lifted and ready beside her.

Instead of backing off, the beast-men kept approaching. They growled and barked guttural noises that didn't sound remotely human.

Primitive terror settled into the back of Annja's mind. During her travels, she'd sometimes come across individuals who lacked mental skills to get by without the kindness of others. But she'd seldom met any of them with violent tendencies. However, she knew that madhouses past and present had been full of such people.

Unable to return to the wall section that had flipped and allowed her access to the hidden passageway, Annja turned and fled along the corridor. Her flashlight beam bounced and jerked as she ran with the beast-men on her tail.

FLEET HEARD the animalistic grunting and roars on the other side of the wall as Patel's demolitions expert wired the door section with shaped plastic explosives. He feared for Annja's life, but there was nothing he could do but wait.

Finally it was done and the demolitions man backed away after sticking radio-controlled detonators into the plastic explosives.

Patel turned to face Fleet. "If she's on the other side of that wall, the flying debris could kill her."

Fleet shook his head and desperately wanted to believe what he said. "She won't be there."

"I hope that you're right." Patel covered his ears, nodded to the demolitions man and turned away.

Fleet plugged his own ears.

A quick series of explosions raced through the amphitheater. Vibrations and the concussive force of the sound filled the large room.

When Fleet looked back at the wall, he saw that a large section of it had been shattered into ruin. She's not going to be happy about that, he thought grimly. But he hoisted himself up and led Patel's men through the opening.

Several bodies of the beast-men blocked the hallway. Thankfully Annja Creed's wasn't one of them. Although his hearing was uncertain and still rang with the sound of the explosions, Fleet heard the sounds of pursuit off to the right.

He picked up the chase and plunged through the darkness.

IN A MATTER OF SECONDS, Annja realized she'd entered a deadly maze of tunnels. Not only that, but the beast-men were herding her in the direction they wanted her to go. Sporadic gunfire blasted through the tunnels. None of it was as loud as the explosion that had taken place behind her.

When she was met by another group of men ahead of her, she entered the tunnel on her left and kept running. Her breath came raggedly, but it was more from the choking dust filling the passageways than because of physical effort.

Only a quick glimpse of the yawning abyss on the other side of the pit in the center of the next room prevented her from plunging over the edge. She came to a stop only inches short of the lip.

Her flashlight beam pushed through most of the darkness. Below, at the bottom of the pit, sharpened wooden stakes, most of them petrified with age and some of them broken from other victims falling and impaling themselves on them, stood more or less upright.

Rajiv Shivaji and his men lay impaled, dead or wounded, among bodies of men and beasts that had fallen prey to the trap over the years. Their flashlights illuminated the gruesome scene.

A few of them still moved. Rajiv was among the living. He stood next to his son, who was impaled through the midsection but still alive.

Annja wouldn't have wished any of them that kind of harm, but she also couldn't erase from her mind the images of the men opening fire on the *Casablanca Moon* only a few days ago.

The surviving men who were able to move were trying to scale the pit wall. Spears in the bodies of other men showed they hadn't been successful and the beast-men had been unmerciful in killing them. Several corpses of the beast-men testified to the fact that the pirates weren't completely helpless.

Growing braver behind her, the beast-men surged forward. They shouted and howled in expectant glee.

Down in the pit, Rajiv Shivaji turned his flashlight on the beast-men and saw Annja. He pulled his rifle to his shoulder and opened fire as she whirled away from the beast-men and the flashlight beam.

The bullets thudded into the beast-men and some of them dropped. In retaliation, the beast-men threw spears and rocks down at the pirates. More of Rajiv's men went down under the onslaught.

Guttural growls and howls erupted from the other side of the room. When she looked across the pit, Annja saw another doorway on the other side. Evidently the room had been designed as a bottleneck for invading forces. The lip around the pit was wide enough to get around, but only in single file. The beast-men navigated it easily, but they didn't have to rush. They could trap their enemies and wait them out.

As Annja watched, the group on the other side of the pit unrolled rope ladders and dropped them down into the pit. They carried spears and swords, and their intent was clear. The pirates redirected their attention to the closer threat, but strategically placed boulders and thick logs provided cover as the beast-men crept among them.

In minutes, Annja knew it would be over and none of them would survive. She stood her ground and lifted the sword. A beast-man carrying a spear came at her and thrust his weapon.

Annja lopped off the front of the spear with her blade, then opened her opponent's throat with the next slash. She spun and kicked the dying man backward. He flailed and took another down with him. They fell onto the pointed stakes below. Annja didn't see them fall, but she heard the crash and splintering wood.

She set herself again. She knew the only chance she had to survive the encounter was to somehow get past the group of beast-men that had followed her down the passageway.

Her desperate fight became a dance of death as she blocked spear and sword thrusts, then returned the fight with her own blows. The beast-men died before her onslaught. In seconds, the tide of battle had turned.

Annja knew she was covered in blood, none of it her own. Her body felt warm and loose, as if she could continue the fight for hours. Gunfire rattled down in the pit and the flashes begged for her attention, but she remained focused on each of the beast-men that stepped out in front of her.

In short order, she was back at the entranceway. Gunfire sounded outside the passageway now. A moment later, she saw muzzle-flashes against the stone walls. As soon as she recognized the potential threat of Patel's men inadvertently shooting her, Annja started to step back out of the way.

When she gave ground, Annja caught sight of a beast-man launching himself at her. She was too late to do anything. The man hit her at the knees and took her down over the edge of the pit.

Annja lost the sword in the wild tumble down the almost straight walls. She knew she could get it back as long as she didn't end up on the pointed end of a stake.

She managed to roll on top of her attacker as they fell

and rode him down the twenty-foot fall. He hit solidly and a stake punched through his chest below his breastbone.

The sudden stop jarred Annja. She gazed down at the stake thrust through the man and knew that if it had slid through another few inches it would have impaled her, as well.

The glow of the flashlights scattered across the bottom of the pit backlit the scene enough for Annja to see most of the details. Annja saw that a pirate had already been impaled on the stake. That body had kept the stake from punching through into Annja, as well.

Lucky, she told herself. She rolled off the dead man and dropped a couple feet to the ground. She willed the sword into her hand and dodged as two beast-men attacked. Both of them had bullet holes stitched across their chests and faces.

A beast-man sprang at her from behind a boulder. He had surprise on his side and he was extremely fast. Annja blocked his overhead sweep with his single-bitted ax and riposted before putting her blade through his heart. When the sword jammed in between the bones, she left it and willed it into her hand a step later. When it appeared, it was clean of blood.

She fought amid the stakes, logs and boulders. When an enemy approached her, she cut him or her down. In minutes, a pile of bodies ringed her position. Men among Patel's group added to the body count.

In the end, though, there were too many of the beast-men. They came on, hardly breaking stride as they grew more certain of their victory.

Annja's breath grew short and her arm grew tired. A

beast-man swung a club at her head. She ducked behind a stake that had a body draped over it, let the stake absorb the attack, then swung around the stake and kicked the man in the back of the head.

"Annja!"

When she heard Fleet's voice, Annja glanced up and saw the man dropping a rope over the edge of the pit.

"Come on!" Fleet urged. "We'll cover you!"

Immediately, Annja altered her course and headed for the rope. A pirate tried to climb it and took a spear through the back. Weakly, he fumbled at the rope and fell backward.

Annja's next step brought her within ten feet of Rajiv Shivaji. The pirate turned toward her and aimed his assault rifle. Annja dodged out of the way, but she knew she couldn't stay where she was. The beast-men were only a few feet away and closing.

"I'm out!" someone yelled.

Annja knew the man was referring to his ammunition. Once the bullets were gone, they wouldn't be able to hold the line against the beast-men.

"I'll offer you a deal, Miss Creed," Rajiv called out.

Annja peered around the corner. Even if Patel's men saw the danger Rajiv presented to her, he was too deeply dug in to be easily reached.

"I'm listening," she said.

"I'll give you your life if you'll give my son his," Rajiv said. "I don't want him to die."

The request surprised Annja. She'd been prepared for Rajiv to try to negotiate his own rescue. Goraksh still hung on the stake. Blood soaked his shirt and pants.

There was no time to think. The confusion in the pit, aided by the darkness, made it hard for Patel's men to lay down covering fire. Not only that, but a few of them were down in the pit now, too. The beast-men were good with the spears.

"I'll do it," Annja agreed. She willed the sword away and stepped out into the open. For a moment, when Rajiv's aim rested on the center of her body, she thought perhaps he was going to shoot her anyway.

But he swiveled around and shot two advancing beast-men. More took their places.

Annja ran to the stake where Goraksh was impaled. Her quick assessment told her that it was possible nothing vital had been hit. But pulling him off the stake was going to open the wound and let it bleed.

She called the sword forth and swung the blade. The keen edge passed right through the stake. She let the weapon fade away. She managed to catch Goraksh over her shoulder before he could fall.

With the slightly built young man over her shoulder, the stake section hard against her back, Annja ran for the rope Fleet had dangling. She grabbed it with her free hand, then pulled her other hand around as she swung her boots against the side of the pit.

She climbed, but she merely touched a couple of times because Fleet and a couple other men pulled her up the side of the pit. Spears and clubs smacked against the wall but missed her for the most part. A club or a rock hit her between the shoulder blades, and a spear grazed her ribs.

When she got to the top of the pit, Patel's men took

Goraksh from her. She turned and took a last look in the pit. The beast-men had swarmed over the pirates and dragged them down. Rajiv fought them with his rifle butt, but he lasted only a few seconds. The beast-men fell on their victims and howled at their victory.

"Come on," Fleet said. "We've burned through nearly all our ammunition. We can't stay here. There are too many of them."

Annja turned and ran. She kept pace with the paratroopers easily. Only a few beast-men tried to block their path. Either they'd decided to let them go because they were too dangerous or they'd all focused on the easier prey down in the pit.

Minutes later, they piled into the boats and sped away.

Annja hated leaving. It must have showed on her face.

"It's not over," Fleet said. "Nobody's going to let go of this site until they know for certain what's here."

"I know," Annja said. "I just want to be part of it."

"If anyone has earned that right, Miss Creed," Patel said grimly, "it's you. The Archaeology Survey of India is not so hard-hearted that they will forget who brought them here."

Annja sincerely hoped so.

epilogue

"They told me I'd find you back here."

At the sound of the familiar voice, Annja turned and saw James Fleet approaching her across the interior courtyard of Kumari Kandam II. The site wasn't actually designated that. It was just something all the workers had taken to calling it.

Fleet looked rested and happy. He gazed around at the site. "You've made a lot of changes."

He was referring to the electrical lighting that hung along the walls and lit up the interior of the subterranean city. Several teams of archaeologists, all of them there by invitation only, labored on different facets of the project.

"I didn't make the changes," Annja said. "The ASI did."

"Ah," Fleet said.

"I can't say that I agree with them all, but I can't take pictures in the dark." Annja held up her digital camera.

"Well, then, modernization in moderation, I suppose," Fleet said.

Annja grinned. "What brings you out all this way?"

"I was curious." Fleet walked around and gazed at the walls. "I hadn't seen it since we left it. I've thought about it a lot. The whole thing has made me curious."

"Curiosity is a natural trait of an archaeologist," Annja said.

"It's also one of an investigator," Fleet said. "I'd get my fill of looking at rock murals pretty quickly, I'm afraid. I'm far more interested in the evils that men do than any works of art they've created."

"They've done evil pretty much since they took their first breaths," Annja said.

"I suppose it all began when Cain slew Abel."

"It probably started even before then. It's just not recorded," Annja said.

"I see the ASI allowed you to work the site."

"They did." Annja picked up a bottle of water from the small refrigerated cooler on the table near where she worked. She offered one to Fleet and he accepted. "It was easy. It was harder finding the time to stay here. I keep a busy schedule."

"I figured that." Fleet sipped the water. "So are you going to be here long?"

"A few weeks."

"Do you plan to eat dinner while you're here?"

Annja smiled. "I eat dinner every night, Agent Fleet."

"Please, call me James. It takes too long to say, 'Yes, Agent Fleet, I'd love to have dinner with you the next time the helicopter takes a group out of here.'"

"I didn't know I was going to say that."

Fleet looked a little rueful. "I was rather hoping that you might."

"All right."

"Good." Fleet nodded.

"But I thought you were busy. That's why you stayed in Kanyakumari."

"Paperwork. It's done now. I've got a staggering caseload, but nothing that can't wait a few days. The bureau is pretty happy knowing Shivaji is off the books. And the .357 Magnum Patel's soldiers found is going to clear a lot of cases."

Annja turned back to the wall where the demolitions man had blown the secret door apart. "I really wish all the damage hadn't been done."

"Couldn't be helped," Fleet replied. "Was much more damage incurred when Patel arrived with reinforcements?"

"No." Annja had had to admire the special-forces team. They'd organized and run efforts to remove the remaining residents of the city.

There had been a few women and children, and the whole thing had been pretty horrible. None of them or the bloodline was going to be salvageable. Almost two hundred survivors were living in old leper compounds outside of Kanyakumari. Medical teams were trying to make them as comfortable as possible, but they were facing certain extinction.

The saddest part was that they had no real oral history to pass on. Their language so far was indecipherable, and most of it was repetitive and nonsense according to the linguistic experts studying them.

Clearing the city had taken most of three days and had included tranquilizer darts and knockout gas. Even then the peaceful solution to the occupancy problem hadn't been totally without loss.

"Mostly it was just sad," Annja admitted. "The military knocked out the residents, loaded them into cargo nets and flew them out of here. Their quality of life isn't going to be very good."

"It'll be safer wherever they are than where they were," Fleet said. "At least there they won't be cannibalizing each other."

"How is Goraksh?" she asked.

"Getting better. It'll still be a while before they release him from the hospital. But I'm told he should make a full recovery."

"That's good."

"I've done some checking around. By all accounts, he's not a hardcore pirate like his father."

Annja nodded. "He wasn't anything like his father."

"In the end, his father wasn't like the man I thought I was chasing," Fleet admitted.

"The pirate who murdered indiscriminately, yet gave his life to save his son?"

"Yes."

"Everyone has family," Annja said. "Those bonds tend to be strong. They either lift people up or they weigh them down. In Shivaji's case, I think he rose to the occasion at the end."

Fleet looked around the courtyard. "They tell me there's no flight back to civilization for another three days."

"That's right," Annja said.

"Three days is a lot of time to do nothing."

"Maybe," Annja said with a smile, "we should find you something to do." She quickly set him to the task of clearing away some of the debris. It all had to be sorted through. If possible, she intended to rebuild the wall. A lot of the history of the people leaving Kumari Kandam was contained on the walls, and she wanted all of it.

Her phone rang as she turned back to her camera. A quick glance at caller ID showed her that it was a New York number.

"Hello," she answered.

"Annja!" Doug Morrell said excitedly. "Long time no hear from my favorite television personality."

Still feeling irritated at the whole memorial DVD collection, Annja leaned against the wall and said nothing.

"Look, I know you're busy," Doug said. "I've been trying to keep up with the crocodile people you found over there."

"They're not crocodile people," Annja said.

Doug sighed. "That's too bad. Crocodile people I could make a show out of."

"They were cannibals," Annja said when she decided to throw him a bone.

"Really? Cool. We can't get enough shows about cannibals."

"This isn't my story," Annja said.

"What do you mean it's not your story? You found the crocodile people." Doug sounded a little outraged.

"This story belongs to the ASI. If I can negotiate some material from them, that'll be fine."

"Work on that. We're running thin on subject material."

"All right."

"Look," Doug went on, "I know you were miffed about that memorial DVD we offered—"

"More than miffed," Annja interrupted.

"—and I came up with an ingenious idea, if I do say so myself, that made the marketing department happy and should make you happy, too."

Annja waited, really not wanting to know.

"Sticker," Doug said.

"Sticker?" Annja didn't see where Doug was headed.

"Sure. We're going to change the DVD to read *Best of* instead of *Memorial* by putting a covering sticker on every DVD case."

Annja couldn't believe it.

"Since we changed the artwork on the page, a lot of people are starting to order it again. That's as much a tribute to you as to anything I could think of."

"Thanks."

"So when are you going to be out of there?"

Annja knew he'd already forgotten where she was. "When I get done."

"If that's going to be a while, we're in trouble," Doug said. "Our production cupboard is empty. We need a story we can run with for the show pretty soon."

Without warning, a yell went up.

"I've gotta go, Doug." Annja closed the phone and vaulted into action. She couldn't help wondering if some of the city's original residents had returned from the jungle where some of them had hidden.

Instead, the archaeology team gathered in the amphitheater stepped back from a hole in the wall. Steps led up inside the hidden passageway.

"There's a library up here," a young man shouted as he paused almost halfway down to his knees on the cut-stone steps.

"A library," Fleet said. "That's important?"

"It is," Annja said, her voice tight with emotion. "None of those people living here were historians. If not for these books, whatever they were before might have been destroyed. It would have been lost forever." She sighed. "Of course, from everything we've seen so far, we can't read their language."

"Eventually someone somewhere will be able to," Fleet said. "Language is a code. There are a lot of people who love code-breaking almost as much as you like chasing after lost things."

"I know."

"I take it they're not going to pass the books around for a while?"

Annja shook her head. "Not until researchers have had a chance to film the room." She couldn't help wondering how large the room was and how many books were inside. "They want to see everything in situ, in case that's important. It generally is."

"Then it sounds like you have time to stretch your legs," Fleet said.

"I should be working."

"You're not going to get any work done. You're going to be thinking about those books."

Annja frowned. "You're probably right."

"Afternoon sunlight outside," Fleet said. "We could stop by the canteen and pick up some food for an impromptu picnic. There are some pretty places around here for a walk."

Annja's first inclination was to turn down the offer. But she'd been working hard. When she hadn't been working on the city site, she'd been thinking about how she was going to finish the excavation on the sunken Roman trade ship. Both sites had things to offer, and she had an opportunity to finish both.

But a walk—just for a while—sounded good.

"All right," she said, and she went with Fleet out into the bright afternoon.

ROOM 59

A research facility in China has built
the ultimate biological weapon. Alex's job:
infiltrate and destroy. His wife works at the
biotech company's stateside lab, and Alex
fears danger is poised to hit home. But when
Alex is captured, his personal and professional
worlds collide in a last, desperate gamble to
stop ruthless masterminds from unleashing
virulent, unstoppable death.

Look for

out of time
by
cliff RYDER

GOLD EAGLE ®

*Available April
wherever you buy books.*

GRM592

TAKE 'EM FREE

2 action-packed novels plus a mystery bonus

NO RISK
NO OBLIGATION TO BUY

Don Pendleton's Mack Bolan®

The Killing Rule

The disappearance of two CIA agents in London plus intelligence involving the IRA and access to weapons of mass destruction launch Bolan's hard probe in the British Isles. Bolan recruits a renegade force to close in on a traitor high in the ranks of the British government. All that stands between a desert continent and a crippling blow to humanity is Bolan's sheer determination to take whatever action is necessary to thwart a victory for terror.

Available January wherever you buy books.